Teacher, Student, and Parent
One-Stop Internet Resources

Log on to
bookl.msscience.com

ONLINE STUDY TOOLS

- Section Self-Check Quizzes
- Interactive Tutor
- Chapter Review Tests
- Standardized Test Practice
- Vocabulary PuzzleMaker

ONLINE RESEARCH

- WebQuest Projects
- Prescreened Web Links
- Career Links
- Internet Labs

INTERACTIVE ONLINE STUDENT EDITION

- Complete Interactive Student Edition available at mhln.com

FOR TEACHERS

- Teacher Bulletin Board
- Teaching Today—Professional Development

SAFETY SYMBOLS

SAFETY SYMBOLS	HAZARD	EXAMPLES	PRECAUTION	REMEDY
DISPOSAL	Special disposal procedures need to be followed.	certain chemicals, living organisms	Do not dispose of these materials in the sink or trash can.	Dispose of wastes as directed by your teacher.
BIOLOGICAL	Organisms or other biological materials that might be harmful to humans	bacteria, fungi, blood, unpreserved tissues, plant materials	Avoid skin contact with these materials. Wear mask or gloves.	Notify your teacher if you suspect contact with material. Wash hands thoroughly.
EXTREME TEMPERATURE	Objects that can burn skin by being too cold or too hot	boiling liquids, hot plates, dry ice, liquid nitrogen	Use proper protection when handling.	Go to your teacher for first aid.
SHARP OBJECT	Use of tools or glassware that can easily puncture or slice skin	razor blades, pins, scalpels, pointed tools, dissecting probes, broken glass	Practice common-sense behavior and follow guidelines for use of the tool.	Go to your teacher for first aid.
FUME	Possible danger to respiratory tract from fumes	ammonia, acetone, nail polish remover, heated sulfur, moth balls	Make sure there is good ventilation. Never smell fumes directly. Wear a mask.	Leave foul area and notify your teacher immediately.
ELECTRICAL	Possible danger from electrical shock or burn	improper grounding, liquid spills, short circuits, exposed wires	Double-check setup with teacher. Check condition of wires and apparatus.	Do not attempt to fix electrical problems. Notify your teacher immediately.
IRRITANT	Substances that can irritate the skin or mucous membranes of the respiratory tract	pollen, moth balls, steel wool, fiberglass, potassium permanganate	Wear dust mask and gloves. Practice extra care when handling these materials.	Go to your teacher for first aid.
CHEMICAL	Chemicals can react with and destroy tissue and other materials	bleaches such as hydrogen peroxide; acids such as sulfuric acid, hydrochloric acid; bases such as ammonia, sodium hydroxide	Wear goggles, gloves, and an apron.	Immediately flush the affected area with water and notify your teacher.
TOXIC	Substance may be poisonous if touched, inhaled, or swallowed.	mercury, many metal compounds, iodine, poinsettia plant parts	Follow your teacher's instructions.	Always wash hands thoroughly after use. Go to your teacher for first aid.
FLAMMABLE	Flammable chemicals may be ignited by open flame, spark, or exposed heat.	alcohol, kerosene, potassium permanganate	Avoid open flames and heat when using flammable chemicals.	Notify your teacher immediately. Use fire safety equipment if applicable.
OPEN FLAME	Open flame in use, may cause fire.	hair, clothing, paper, synthetic materials	Tie back hair and loose clothing. Follow teacher's instruction on lighting and extinguishing flames.	Notify your teacher immediately. Use fire safety equipment if applicable.

 Eye Safety Proper eye protection should be worn at all times by anyone performing or observing science activities.

 Clothing Protection This symbol appears when substances could stain or burn clothing.

 Animal Safety This symbol appears when safety of animals and students must be ensured.

 Handwashing After the lab, wash hands with soap and water before removing goggles.

Glencoe Science

Chemistry

NATIONAL GEOGRAPHIC

book1.msscience.com

 **Glencoe**

New York, New York   Columbus, Ohio   Chicago, Illinois   Peoria, Illinois   Woodland Hills, California

Glencoe Science

Chemistry

Water droplets condense on the surface of a flask that has been cooled by a chemical reaction. A chemical change produces new substances that have properties different from those of the original substances. The change in temperature is an indication that a chemical reaction has taken place.

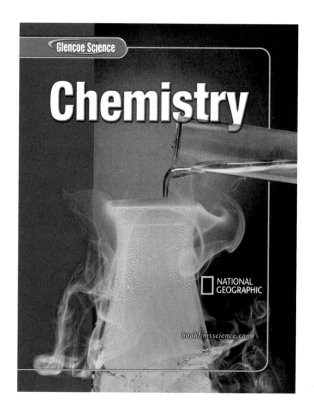

NATIONAL GEOGRAPHIC

book1.msscience.com

Glencoe

The *McGraw-Hill* Companies

Send all inquiries to:
Glencoe/McGraw-Hill
8787 Orion Place
Columbus, OH 43240-4027

ISBN: 0-07-861767-7

Printed in the United States of America.

4 5 6 7 8 9 10 027/111 09 08 07 06

2015 -110

Authors

NATIONAL GEOGRAPHIC
Education Division
Washington, D.C.

Eric Werwa, PhD
Department of Physics and
Astronomy
Otterbein College
Westerville, OH

Dinah Zike
Educational Consultant
Dinah-Might Activities, Inc.
San Antonio, TX

Series Consultants

CONTENT

Linda McGaw
Science Program Coordinator
Advanced Placement Strategies, Inc.
Dallas, TX

MATH

Michael Hopper, DEng
Manager of Aircraft Certification
L-3 Communications
Greenville, TX

READING

Rachel Swaters-Kissinger
Science Teacher
John Boise Middle School
Warsaw, MO

SAFETY

Aileen Duc, PhD
Science 8 Teacher
Hendrick Middle School, Plano ISD
Plano, TX

Sandra West, PhD
Department of Biology
Texas State University-San Marcos
San Marcos, TX

ACTIVITY TESTERS

Nerma Coats Henderson
Pickerington Lakeview Jr. High
School
Pickerington, OH

Mary Helen Mariscal-Cholka
William D. Slider Middle School
El Paso, TX

**Science Kit and Boreal
Laboratories**
Tonawanda, NY

Series Reviewers

Sharla Adams
IPC Teacher
Allen High School
Allen, TX

Desiree Bishop
Environmental Studies Center
Mobile County Public Schools
Mobile, AL

Tom Bright
Concord High School
Charlotte, NC

Joanne Davis
Murphy High School
Murphy, NC

Annette Parrott
Lakeside High School
Atlanta, GA

Nora M. Prestinari Burchett
Saint Luke School
McLean, VA

Karen Watkins
Perry Meridian Middle School
Indianapolis, IN

HOW TO...

Use Your Science Book

Before You Read

- **Chapter Opener** Science is occurring all around you, and the opening photo of each chapter will preview the science you will be learning about. The **Chapter Preview** will give you an idea of what you will be learning about, and you can try the **Launch Lab** to help get your brain headed in the right direction. The **Foldables** exercise is a fun way to keep you organized.

- **Section Opener** Chapters are divided into two to four sections. The **As You Read** in the margin of the first page of each section will let you know what is most important in the section. It is divided into four parts. **What You'll Learn** will tell you the major topics you will be covering. **Why It's Important** will remind you why you are studying this in the first place! The **Review Vocabulary** word is a word you already know, either from your science studies or your prior knowledge. The **New Vocabulary** words are words that you need to learn to understand this section. These words will be in **boldfaced** print and highlighted in the section. Make a note to yourself to recognize these words as you are reading the section.

Glencoe Science

Chemistry

NATIONAL
GEOGRAPHIC

As You Read

- **Headings** Each section has a title in large red letters, and is further divided into blue titles and small red titles at the beginnings of some paragraphs. To help you study, make an outline of the headings and subheadings.

- **Margins** In the margins of your text, you will find many helpful resources. The **Science Online** exercises and **Integrate** activities help you explore the topics you are studying. **MiniLabs** reinforce the science concepts you have learned.

- **Building Skills** You also will find an **Applying Math** or **Applying Science** activity in each chapter. This gives you extra practice using your new knowledge, and helps prepare you for standardized tests.

- **Student Resources** At the end of the book you will find **Student Resources** to help you throughout your studies. These include **Science, Technology,** and **Math Skill Handbooks,** an **English/Spanish Glossary,** and an **Index.** Also, use your **Foldables** as a resource. It will help you organize information, and review before a test.

- **In Class** Remember, you can always ask your teacher to explain anything you don't understand.

FOLDABLES™
Study Organizer

Science Vocabulary Make the following Foldable to help you understand the vocabulary terms in this chapter.

STEP 1 Fold a vertical sheet of notebook paper from side to side.

STEP 2 Cut along every third line of only the top layer to form tabs.

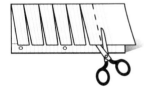

STEP 3 Label each tab with a vocabulary word from the chapter.

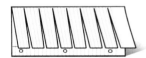

Build Vocabulary As you read the chapter, list the vocabulary words on the tabs. As you learn the definitions, write them under the tab for each vocabulary word.

Look For...

FOLDABLES™

At the beginning of every section.

In Lab

Working in the laboratory is one of the best ways to understand the concepts you are studying. Your book will be your guide through your laboratory experiences, and help you begin to think like a scientist. In it, you not only will find the steps necessary to follow the investigations, but you also will find helpful tips to make the most of your time.

● Each lab provides you with a **Real-World Question** to remind you that science is something you use every day, not just in class. This may lead to many more questions about how things happen in your world.

● Remember, experiments do not always produce the result you expect. Scientists have made many discoveries based on investigations with unexpected results. You can try the experiment again to make sure your results were accurate, or perhaps form a new hypothesis to test.

● Keeping a **Science Journal** is how scientists keep accurate records of observations and data. In your journal, you also can write any questions that may arise during your investigation. This is a great method of reminding yourself to find the answers later.

Look For...
● **Launch Labs** start every chapter.
● **MiniLabs** in the margin of each chapter.
● **Two Full-Period Labs** in every chapter.
● **EXTRA Try at Home Labs** at the end of your book.
● the **Web site** with laboratory demonstrations.

Before a Test

Admit it! You don't like to take tests! However, there *are* ways to review that make them less painful. Your book will help you be more successful taking tests if you use the resources provided to you.

- Review all of the **New Vocabulary** words and be sure you understand their definitions.

- Review the notes you've taken on your **Foldables,** in class, and in lab. Write down any question that you still need answered.

- Review the **Summaries** and **Self Check questions** at the end of each section.

- Study the concepts presented in the chapter by reading the **Study Guide** and answering the questions in the **Chapter Review.**

Look For...

- Reading Checks and caption questions throughout the text.
- the Summaries and Self Check questions at the end of each section.
- the Study Guide and Review at the end of each chapter.
- the Standardized Test Practice after each chapter.

Let's Get Started

To help you find the information you need quickly, use the Scavenger Hunt below to learn where things are located in Chapter 1.

1. What is the title of this chapter?

2. What will you learn in Section 1?

3. Sometimes you may ask, "Why am I learning this?" State a reason why the concepts from Section 2 are important.

4. What is the main topic presented in Section 2?

5. How many reading checks are in Section 1?

6. What is the Web address where you can find extra information?

7. What is the main heading above the sixth paragraph in Section 2?

8. There is an integration with another subject mentioned in one of the margins of the chapter. What subject is it?

9. List the new vocabulary words presented in Section 2.

10. List the safety symbols presented in the first Lab.

11. Where would you find a Self Check to be sure you understand the section?

12. Suppose you're doing the Self Check and you have a question about concept mapping. Where could you find help?

13. On what pages are the Chapter Study Guide and Chapter Review?

14. Look in the Table of Contents to find out on which page Section 2 of the chapter begins.

15. You complete the Chapter Review to study for your chapter test. Where could you find another quiz for more practice?

Contents

In each chapter, look for these opportunities for review and assessment:
- Reading Checks
- Caption Questions
- Section Review
- Chapter Study Guide
- Chapter Review
- Standardized Test Practice
- Online practice at **book1.msscience.com**

Student Resources

Cross-Curricular Readings/Labs

available as a video lab

Content Details

INTEGRATE

Science Online

Standardized Test Practice

Content Details

Alfred Nobel, Dynamite, and Peace

Figure 1 Alfred Nobel (1833–1896) invented both dynamite and the Nobel prize.

Alfred Nobel is best known for the invention of dynamite and the establishment of the Nobel prize—an award given to those whose efforts in the fields of physics, chemistry, medicine, literature, economics, or peace have benefited humanity. These two seemingly opposite acts—the invention of a deadly explosive and the establishment of a prize that promotes, among other things, peace—emphasize the power and the limitations of science.

Science enabled Nobel to create a superior explosive, but science could not answer other important questions, such as how dynamite should be used. Was it ethical for Nobel to invent an explosive that increased the killing power of weapons? Are scientists responsible for how their discoveries are used? Perhaps Nobel's own answer to these questions was to bestow money after his death for the establishment of the Nobel prize.

A Powerful Invention

In the 1850s, Nobel began experimenting with nitroglycerin (ni troh GLIHS or ohn)—a powerful liquid explosive. Since it exploded unpredictably, it was considered too dangerous for widespread use. Nobel decided to find a way to make nitroglycerin safer to handle. Nobel called his invention dynamite.

Figure 2 Dynamite can be used to clear away sections of mountains in order to build tunnels for roads and trains.

Nobel intended dynamite to be used as a construction tool. It was more powerful than gunpowder—the most common explosive used in construction at the time. Indeed, dynamite helped reduce the cost of blasting rocks, which is essential for building tunnels and canals. However, military leaders were also interested in Nobel's discovery. Only a few years after its invention, dynamite was used as a weapon in the Franco-Prussian War (1870–1871), a conflict between France and the German state of Prussia.

Nobel had not intended dynamite to be used as a weapon. Still, he became rich from selling dynamite to armies as well as to construction companies. Later he invented other explosives specifically for use in missiles, torpedoes, and cannons.

Some biographers claim that Nobel believed scientists are not responsible for how their discoveries are used. Others assert that Nobel founded a prize that promotes peace to counteract the harm done by his contribution to weapons. Clearly, science can't answer all the questions about scientists's accountability for their discoveries.

Figure 3 Dynamite was originally intended to be used in construction. This tunnel was created with dynamite.

Figure 4 During the Franco-Prussian War (1870–1871), dynamite was used as a weapon.

Science

Science is the process of gaining knowledge by asking questions and seeking the answers to these questions. To answer questions, scientists use scientific methods. They include identifying a question, forming and testing a hypothesis, analyzing results, and drawing conclusions. Alfred Nobel invented dynamite by beginning with the question "How can nitroglycerin be made more stable and therefore safer to handle?" After forming a hypothesis that nitroglycerin would be more stable if mixed with another substance, he tested several materials and finally found a safer mixture. For a question to be scientific, it must be testable. Scientific conclusions can change as more information is gained.

The Power of Science

Alfred Nobel's invention has benefited humankind in countless ways. The Panama Canal, Mount Rushmore, and many tunnels and mines were built with the aid of dynamite. It can break up dangerous ice and logjams and it can quickly and safely reduce large buildings to rubble. Police departments use dynamite to detonate suspicious packages. Fire departments use it to put out oil well fires. The explosion of the dynamite requires a huge amount of oxygen and suffocates the fire.

Figure 4 Before there were faces on Mount Rushmore, there was unshaped rock. About 90 percent of the mountain was carved with dynamite.

Figure 5 In demolitions, explosions are carefully placed to ensure that the building collapses inward.

The Limits of Science

Using scientific methods is the best way to learn about how the world works, but science has its limitations. Scientists are sometimes unable to answer a question or solve a problem because they lack the necessary tools. This is often a temporary limitation because once the required tools are developed, science often provides answers. For example, scientists were unaware of the existence of Jupiter's moons before the telescope was invented.

What Science Can't Answer

Science can't give answers to questions that are not testable or that can't be measured or observed. Three major areas in which science can't provide answers are questions about morality, values, and spirituality.

The idea that it was immoral for Nobel to sell dynamite to armies, for example, is not a scientific idea because it can't be scientifically tested. Similarly, science can't answer opinion questions about values like the modern-day question of how far advances in the field of cloning should be taken. Should scientists use the new techniques developed to clone a human being? Any possible answer is a matter of opinion and therefore can't be measured or tested. Finally, science can't answer questions about spiritual matters because they are unable to be observed, measured, or tested.

Science and Responsibility

Although science can't answer questions about ethics and values, it can provide facts that may help people to make informed decisions. Being familiar with facts on all sides of an issue and careful consideration of what the possible positive and negative effects might be to an individual, a society, or the environment can help people decide upon a course of action.

Figure 6 When the world was presented with Dolly, a sheep produced as a result of cloning, many ethical questions were raised about future applications of this new biotechnology.

You Do It

Each new scientific discovery brings new questions. Some of these questions concern ethical matters that can't be answered by science alone. Find out about a recipient of the Nobel Prize for Chemistry. What did the person do to win this honor? What ethical questions arise from his or her work?

Atomic Structure and Chemical Bonds

The Noble Family

Blimps, city lights, and billboards, all have something in common—they use gases that are members of the same element family. In this chapter, you'll learn about the unique properties of element families. You'll also learn how electrons can be lost, gained, and shared by atoms to form chemical bonds.

Science Journal Write a sentence comparing household glue to chemical bonds.

Start-Up Activities

Model the Energy of Electrons

It's time to clean out your room—again. Where do all these things come from? Some are made of cloth and some of wood. The books are made of paper and an endless array of things are made of plastic. Fewer than 100 different kinds of naturally occurring elements are found on Earth. They combine to make all these different substances. What makes elements form chemical bonds with other elements? The answer is in their electrons.

1. Pick up a paper clip with a magnet. Touch that paper clip to another paper clip and pick it up.

2. Continue picking up paper clips this way until you have a strand of them and no more will attach.

3. Then, gently pull off the paper clips one by one.

4. **Think Critically** In your Science Journal, discuss which paper clip was easiest to remove and which was hardest. Was the clip that was easiest to remove closer to or farther from the magnet?

Study Organizer

Chemical Bonds Make the following Foldable to help you classify information by diagramming ideas about chemical bonds.

STEP 1 Fold a vertical sheet of paper in half from top to bottom.

STEP 2 Fold in half from side to side with the fold at the top.

STEP 3 Unfold the paper once. Cut only the fold of the top flap to make two tabs.

STEP 4 Turn the paper vertically and label the tabs as shown.

Ionic Bonds

Covalent Bonds

Summarize As you read the chapter, identify the main ideas of bonding under the appropriate tabs. After you have read the chapter, explain the difference between polar covalent bonds and covalent bonds on the inside portion of your Foldable.

Preview this chapter's content and activities at
bookl.msscience.com

Why do atoms combine?

What **You'll Learn**

- **Identify** how electrons are arranged in an atom.
- **Compare** the relative amounts of energy of electrons in an atom.
- **Compare** how the arrangement of electrons in an atom is related to its place in the periodic table.

Why **It's Important**

Chemical reactions take place all around you.

🔍 **Review Vocabulary**

atom: the smallest part of an element that keeps all the properties of that element

New Vocabulary
- electron cloud
- energy level
- electron dot diagram
- chemical bond

Figure 1 You can compare and contrast electrons with planets.

Atomic Structure

When you look at your desk, you probably see it as something solid. You might be surprised to learn that all matter, even solids like wood and metal contain mostly empty space. How can this be? The answer is that although there might be little or no space between atoms, a lot of empty space lies within each atom.

At the center of every atom is a nucleus containing protons and neutrons. This nucleus represents most of the atom's mass. The rest of the atom is empty except for the atom's electrons, which are extremely small compared with the nucleus. Although the exact location of any one electron cannot be determined, the atom's electrons travel in an area of space around the nucleus called the **electron cloud.**

To visualize an atom, picture the nucleus as the size of a penny. In this case, electrons would be smaller than grains of dust and the electron cloud would extend outward as far as 20 football fields.

Electrons You might think that electrons resemble planets circling the Sun, but they are very different, as you can see in **Figure 1.** First, planets have no charges, but the nucleus of an atom has a positive charge and electrons have negative charges.

Second, planets travel in predictable orbits—you can calculate exactly where one will be at any time. This is not true for electrons. Although electrons do travel in predictable areas, it is impossible to calculate the exact position of any one electron. Instead scientists use a mathematical model that predicts where an electron is most likely to be.

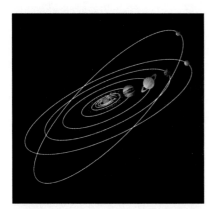

Planets travel in well-defined paths.

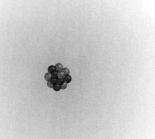

Electrons travel around the nucleus. However, their paths are not well-defined.

Element Structure Each element has a different atomic structure consisting of a specific number of protons, neutrons, and electrons. The number of protons and electrons is always the same for a neutral atom of a given element. **Figure 2** shows a two-dimensional model of the electron structure of a lithium atom, which has three protons and four neutrons in its nucleus, and three electrons moving around its nucleus.

Electron Arrangement

The number and arrangement of electrons in the electron cloud of an atom are responsible for many of the physical and chemical properties of that element.

Electron Energy Although all the electrons in an atom are somewhere in the electron cloud, some electrons are closer to the nucleus than others. The different areas for an electron in an atom are called **energy levels. Figure 3** shows a model of what these energy levels might look like. Each level represents a different amount of energy.

Number of Electrons Each energy level can hold a maximum number of electrons. The farther an energy level is from the nucleus, the more electrons it can hold. The first energy level, energy level 1, can hold one or two electrons, the second, energy level 2, can hold up to eight, the third can hold up to 18, and the fourth energy level can hold a maximum of 32 electrons.

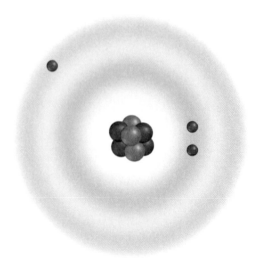

Figure 2 This neutral lithium atom has three positively charged protons, three negatively charged electrons, and four neutral neutrons.

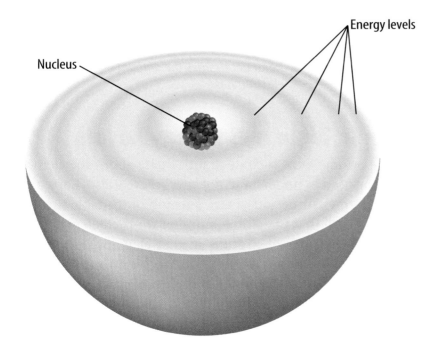

Nucleus

Energy levels

Figure 3 Electrons travel in three dimensions around the nucleus of an atom. The dark bands in this diagram show the energy levels where electrons are most likely to be found.
Identify *the energy level that can hold the most electrons.*

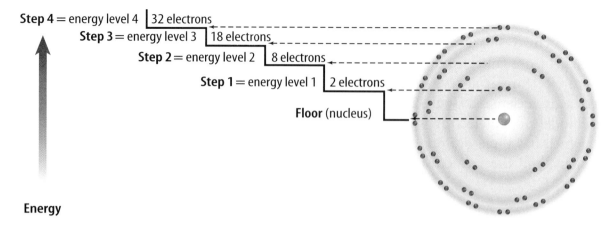

Step 4 = energy level 4 | 32 electrons
Step 3 = energy level 3 | 18 electrons
Step 2 = energy level 2 | 8 electrons
Step 1 = energy level 1 | 2 electrons
Floor (nucleus)

Energy

Figure 4 The farther an energy level is from the nucleus, the more electrons it can hold.
Identify *the energy level with the least energy and the energy level with the most energy.*

Energy Steps The stairway, shown in **Figure 4,** is a model that shows the maximum number of electrons each energy level can hold in the electron cloud. Think of the nucleus as being at floor level. Electrons within an atom have different amounts of energy, represented by energy levels. These energy levels are represented by the stairsteps in **Figure 4.** Electrons in the level closest to the nucleus have the lowest amount of energy and are said to be in energy level one. Electrons farthest from the nucleus have the highest amount of energy and are the easiest to remove. To determine the maximum number of electrons that can occupy an energy level, use the formula, $2n^2$, where n equals the number of the energy level.

Recall the Launch Lab at the beginning of the chapter. It took more energy to remove the paper clip that was closest to the magnet than it took to remove the one that was farthest away. That's because the closer a paper clip was to the magnet, the stronger the magnet's attractive force was on the clip. Similarly, the closer a negatively charged electron is to the positively charged nucleus, the more strongly it is attracted to the nucleus. Therefore, removing electrons that are close to the nucleus takes more energy than removing those that are farther away from the nucleus.

 **Reading Check** *What determines the amount of energy an electron has?*

Topic: Electrons
Visit bookl.msscience.com for Web links to information about electrons and their history.

Activity Research why scientists cannot locate the exact positions of an electron.

Periodic Table and Energy Levels

The periodic table includes a lot of data about the elements and can be used to understand the energy levels also. Look at the horizontal rows, or periods, in the portion of the table shown in **Figure 5.** Recall that the atomic number for each element is the same as the number of protons in that element and that the number of protons equals the number of electrons because an atom is electrically neutral. Therefore, you can determine the number of electrons in an atom by looking at the atomic number written above each element symbol.

Electron Configurations

If you look at the periodic table shown in **Figure 5,** you can see that the elements are arranged in a specific order. The number of electrons in a neutral atom of the element increases by one from left to right across a period. For example, the first period consists of hydrogen with one electron and helium with two electrons in energy level one. Recall from **Figure 4** that energy level one can hold up to two electrons. Therefore, helium's outer energy level is complete. Atoms with a complete outer energy level are stable. Therefore, helium is stable.

✔ Reading Check *What term is given to the rows of the periodic table?*

The second period begins with lithium, which has three electrons—two in energy level one and one in energy level two. Lithium has one electron in its outer energy level. To the right of lithium is beryllium with two outer-level electrons, boron with three, and so on until you reach neon with eight.

Look again at **Figure 4.** You'll see that energy level two can hold up to eight electrons. Not only does neon have a complete outer energy level, but also this configuration of exactly eight electrons in an outer energy level is stable. Therefore, neon is stable. The third period elements fill their outer energy levels in the same manner, ending with argon. Although energy level three can hold up to 18 electrons, argon has eight electrons in its outer energy level—a stable configuration. Each period in the periodic table ends with a stable element.

INTEGRATE
Career

Nobel Prize Winner
Ahmed H. Zewail is a professor of chemistry and physics and the director of the Laboratory for Molecular Sciences at the California Institute of Technology. He was awarded the 1999 Nobel Prize in Chemistry for his research. Zewail and his research team use lasers to record the making and breaking of chemical bonds.

Figure 5 This portion of the periodic table shows the electron configurations of some elements. Count the electrons in each element and notice how the number increases across a period.

1							18
Hydrogen 1 **H**							Helium 2 **He**
	2	13	14	15	16	17	
Lithium 3 **Li**	Beryllium 4 **Be**	Boron 5 **B**	Carbon 6 **C**	Nitrogen 7 **N**	Oxygen 8 **O**	Fluorine 9 **F**	Neon 10 **Ne**
Sodium 11 **Na**	Magnesium 12 **Mg**	Aluminum 13 **Al**	Silicon 14 **Si**	Phosphorus 15 **P**	Sulfur 16 **S**	Chlorine 17 **Cl**	Argon 18 **Ar**

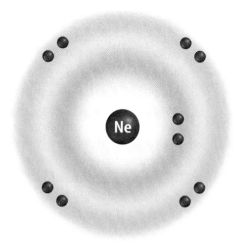

Figure 6 The noble gases are stable elements because their outer energy levels are complete or have a stable configuration of eight electrons like neon shown here.

Figure 7 The halogen element fluorine has seven electrons in its outer energy level.
Determine *how many electrons the halogen family member bromine has in its outer energy level.*

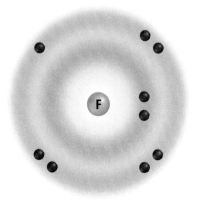

Element Families

Elements can be divided into groups, or families. Each column of the periodic table in **Figure 5** contains one element family. Hydrogen is usually considered separately, so the first element family begins with lithium and sodium in the first column. The second family starts with beryllium and magnesium in the second column, and so on. Just as human family members often have similar looks and traits, members of element families have similar chemical properties because they have the same number of electrons in their outer energy levels.

It was the repeating pattern of properties that gave Russian chemist Dmitri Mendeleev the idea for his first periodic table in 1869. While listening to his family play music, he noticed how the melody repeated with increasing complexity. He saw a similar repeating pattern in the elements and immediately wrote down a version of the periodic table that looks much as it does today.

Noble Gases Look at the structure of neon in **Figure 6.** Neon and the elements below it in Group 18 have eight electrons in their outer energy levels. Their energy levels are stable, so they do not combine easily with other elements. Helium, with two electrons in its lone energy level, is also stable. At one time these elements were thought to be completely unreactive, and therefore became known as the inert gases. When chemists learned that some of these gases can react, their name was changed to noble gases. They are still the most stable element group.

This stability makes possible one widespread use of the noble gases—to protect filaments in lightbulbs. Another use of noble gases is to produce colored light in signs. If an electric current is passed through them they emit light of various colors—orange-red from neon, lavender from argon, and yellowish-white from helium.

Halogens The elements in Group 17 are called the halogens. A model of the element fluorine in period 2 is shown in **Figure 7.** Like all members of this family, fluorine needs one electron to obtain a stable outer energy level. The easier it is for a halogen to gain this electron to form a bond, the more reactive it is. Fluorine is the most reactive of the halogens because its outer energy level is closest to the nucleus. The reactivity of the halogens decreases down the group as the outer energy levels of each element's atoms get farther from the nucleus. Therefore, bromine in period 4 is less reactive than fluorine in period 2.

Alkali Metals Look at the element family in Group 1 on the periodic table at the back of this book, called the alkali metals. The first members of this family, lithium and sodium, have one electron in their outer energy levels. You can see in **Figure 8** that potassium also has one electron in its outer level. Therefore, you can predict that the next family member, rubidium, does also. These electron arrangements are what determines how these metals react.

Reading Check *How many electrons do the alkali metals have in their outer energy levels?*

The alkali metals form compounds that are similar to each other. Alkali metals each have one outer energy level electron. It is this electron that is removed when alkali metals react. The easier it is to remove an electron, the more reactive the atom is. Unlike halogens, the reactivities of alkali metals increase down the group; that is, elements in the higher numbered periods are more reactive than elements in the lower numbered periods. This is because their outer energy levels are farther from the nucleus. Less energy is needed to remove an electron from an energy level that is farther from the nucleus than to remove one from an energy level that is closer to the nucleus. For this reason, cesium in period 6 loses an electron more readily and is more reactive than sodium in period 3.

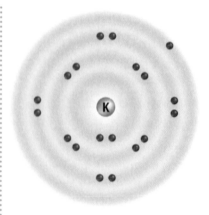

Figure 8 Potassium, like lithium and sodium, has only one electron in its outer level.

Applying Science

How does the periodic table help you identify properties of elements?

The periodic table displays information about the atomic structure of the elements. This information includes the properties, such as the energy level, of the elements. Can you identify an element if you are given information about its energy level? Use your ability to interpret the periodic table to find out.

Identifying the Problem

Recall that elements in a group in the periodic table contain the same number of electrons in their outer levels. The number of electrons increases by one from left to right across a period. Refer to **Figure 5.** Can you identify an

unknown element or the group a known element belongs to?

Solving the Problem

1. An unknown element in Group 2 has a total number of 12 electrons and two electrons in its outer level. What is it?
2. Name the element that has eight electrons, six of which are in its outer level.
3. Silicon has a total of 14 electrons, four electrons in its outer level, and three energy levels. What group does silicon belong to?
4. Three elements have the same number of electrons in their outer energy levels. One is oxygen. Using the periodic table, what might the other two be?

Electron Dot Diagrams

You have read that the number of electrons in the outer energy level of an atom determines many of the chemical properties of the atom. Because these electrons are so important in determining the chemical properties of atoms, it can be helpful to make a model of an atom that shows only the outer electrons. A model like this can be used to show what happens to these electrons during reactions.

Drawing pictures of the energy levels and electrons in them takes time, especially when a large number of electrons are present. If you want to see how atoms of one element will react, it is handy to have an easier way to represent the atoms and the electrons in their outer energy levels. You can do this with electron dot diagrams. An **electron dot diagram** is the symbol for the element surrounded by as many dots as there are electrons in its outer energy level. Only the outer energy level electrons are shown because these are what determine how an element can react.

How to Write Them How do you know how many dots to make? For Groups 1 and 2, and 13–18, you can use the periodic table or the portion of it shown in **Figure 5**. Group 1 has one outer electron. Group 2 has two. Group 13 has three, Group 14, four, and so on to Group 18. All members of Group 18 have stable outer energy levels. From neon down, they have eight electrons. Helium has only two electrons, because that is all that its single energy level can hold.

The dots are written in pairs on four sides of the element symbol. Start by writing one dot on the top of the element symbol, then work your way around, adding dots to the right, bottom, and left. Add a fifth dot to the top to make a pair. Continue in this manner until you reach eight dots to complete the level.

The process can be demonstrated by writing the electron dot diagram for the element nitrogen. First, write N—the element symbol for nitrogen. Then, find nitrogen in the periodic table and see what group it is in. It's in Group 15, so it has five electrons in its outer energy level. The completed electron dot diagram for nitrogen can be seen in **Figure 9**.

The electron dot diagram for iodine can be drawn the same way. The completed diagram is shown on the right in **Figure 9**.

Figure 9 Electron dot diagrams show only the electrons in the outer energy level.
Explain why only the outer energy level electrons are shown.

Nitrogen contains five electrons in its outer energy level.

Iodine contains seven electrons in its outer energy level.

Figure 10 Some models are made by gluing pieces together. The glue that holds elements together in a chemical compound is the chemical bond.

Using Dot Diagrams Now that you know how to write electron dot diagrams for elements, you can use them to show how atoms bond with each other. A **chemical bond** is the force that holds two atoms together. Chemical bonds unite atoms in a compound much as glue unites the pieces of the model in **Figure 10.** Atoms bond with other atoms in such a way that each atom becomes more stable. That is, their outer energy levels will resemble those of the noble gases.

✔ **Reading Check** *What is a chemical bond?*

section 1 review

Summary

Atom Structure
- At the center of the atom is the nucleus.
- Electrons exist in an area called the electron cloud.
- Electrons have a negative charge.

Electron Arrangement
- The different regions for an electron in an atom are called energy levels.
- Each energy level can hold a maximum number of electrons.

The Periodic Table
- The number of electrons is equal to the atomic number.
- The number of electrons in a neutral atom increases by one from left to right across a period.

Self Check

1. **Determine** how many electrons nitrogen has in its outer energy level. How many does bromine have?

2. **Solve** for the number of electrons that oxygen has in its first energy level. Second energy level?

3. **Identify** which electrons in oxygen have more energy, those in the first energy level or those in the second.

4. **Think Critically** Atoms in a group of elements increase in size as you move down the columns in the periodic table. Explain why this is so.

Applying Math

5. **Solve One-Step Equations** You can calculate the maximum number of electrons each energy level can hold using the formula $2n^2$. Calculate the number of electrons in the first five energy levels where n equals the number of energy levels.

section 2

How Elements Bond

Ionic Bonds—Loss and Gain

When you put together the pieces of a jigsaw puzzle, they stay together only as long as you wish. When you pick up the completed puzzle, it falls apart. When elements are joined by chemical bonds, they do not readily fall apart. What would happen if suddenly the salt you were shaking on your fries separated into sodium and chlorine? Atoms form bonds with other atoms using the electrons in their outer energy levels. They have four ways to do this—by losing electrons, by gaining electrons, by pooling electrons, or by sharing electrons with another element.

Sodium is a soft, silvery metal as shown in **Figure 11.** It can react violently when added to water or to chlorine. What makes sodium so reactive? If you look at a diagram of its energy levels below, you will see that sodium has only one electron in its outer level. Removing this electron empties this level and leaves the completed level below. By removing one electron, sodium's electron configuration becomes the same as that of the stable noble gas neon.

Chlorine forms bonds in a way that is the opposite of sodium—it gains an electron. When chlorine accepts an electron, its electron configuration becomes the same as that of the noble gas argon.

Figure 11 Sodium and chlorine react, forming white crystalline sodium chloride.

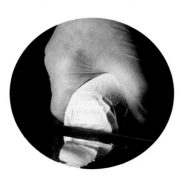

Sodium

Sodium is a silvery metal that can be cut with a knife. Chlorine is a greenish, poisonous gas.

Chlorine

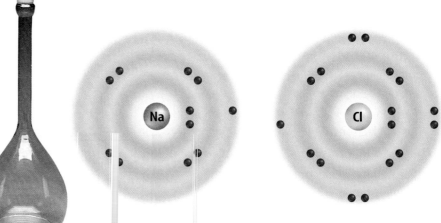

Their electronic structures show why they react.

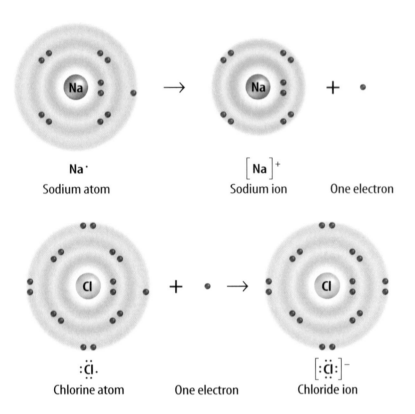

Na·
Sodium atom

$\left[\text{Na}\right]^+$
Sodium ion

One electron

:Ċl·
Chlorine atom

One electron

$\left[:\ddot{\text{Cl}}:\right]^-$
Chloride ion

Figure 12 Ions form when elements lose or gain electrons. When sodium comes into contact with chlorine, an electron is transferred from the sodium atom to the chlorine atom. Na becomes a Na$^+$ ion. Cl becomes a Cl$^-$ ion.

Ions When ions dissolve in water, they separate. Because of their positive and negative charges, the ions can conduct an electric current. If wires are placed in such a solution and the ends of the wires are connected to a battery, the positive ions move toward the negative terminal and the negative ions move toward the positive terminal. This flow of ions completes the circuit.

Ions—A Question of Balance As you just learned, a sodium atom loses an electron and becomes more stable. But something else happens also. By losing an electron, the balance of electric charges changes. Sodium becomes a positively charged ion because there is now one fewer electron than there are protons in the nucleus. In contrast, chlorine becomes an ion by gaining an electron. It becomes negatively charged because there is one more electron than there are protons in the nucleus.

An atom that is no longer neutral because it has lost or gained an electron is called an **ion** (I ahn). A sodium ion is represented by the symbol Na$^+$ and a chloride ion is represented by the symbol Cl$^-$. **Figure 12** shows how each atom becomes an ion.

Bond Formation The positive sodium ion and the negative chloride ion are strongly attracted to each other. This attraction, which holds the ions close together, is a type of chemical bond called an **ionic bond.** In **Figure 13,** sodium and chloride ions form an ionic bond. The compound sodium chloride, or table salt, is formed. A **compound** is a pure substance containing two or more elements that are chemically bonded.

$$\text{Na}° \quad + \quad ·\ddot{\text{Cl}}: \quad \rightarrow \quad \left[\text{Na}\right]^+ \left[:\ddot{\text{Cl}}:\right]^-$$

Figure 13 An ionic bond forms between atoms of opposite charges. **Describe** *how an atom becomes positive or negative.*

Figure 14 Magnesium has two electrons in its outer energy level.

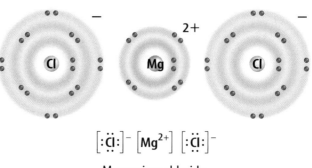

If one electron is lost to each of two chlorine atoms, magnesium chloride forms.

$$\left[:\ddot{Cl}:\right]^{-} \left[Mg^{2+}\right] \left[:\ddot{Cl}:\right]^{-}$$

Magnesium chloride

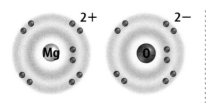

$$\left[Mg^{2+}\right] \left[O^{2-}\right]$$

Magnesium oxide

If both electrons are lost to one oxygen atom, magnesium oxide forms.

Determine *the electron arrangement for magnesium sulfide and calcium oxide.*

More Gains and Losses You have seen what happens when elements gain or lose one electron, but can elements lose or gain more than one electron? The element magnesium, Mg, in Group 2 has two electrons in its outer energy level. Magnesium can lose these two electrons and achieve a completed energy level. These two electrons can be gained by two chlorine atoms. As shown in **Figure 14,** a single magnesium ion represented by the symbol Mg^{2+} and two chloride ions are generated. The two negatively charged chloride ions are attracted to the positively charged magnesium ion forming ionic bonds. As a result of these bonds, the compound magnesium chloride ($MgCl_2$) is produced.

Some atoms, such as oxygen, need to gain two electrons to achieve stability. The two electrons released by one magnesium atom could be gained by a single atom of oxygen. When this happens, magnesium oxide (MgO) is formed, as shown in **Figure 14.** Oxygen can form similar compounds with any positive ion from Group 2.

Metallic Bonding—Pooling

You have just seen how metal atoms form ionic bonds with atoms of nonmetals. Metals can form bonds with other metal atoms, but in a different way. In a metal, the electrons in the outer energy levels of the atoms are not held tightly to individual atoms. Instead, they move freely among all the ions in the metal, forming a shared pool of electrons, as shown in **Figure 15. Metallic bonds** form when metal atoms share their pooled electrons. This bonding affects the properties of metals. For example, when a metal is hammered into sheets or drawn into a wire, it does not break. Instead, layers of atoms slide over one another. The pooled electrons tend to hold the atoms together. Metallic bonding also is the reason that metals conduct electricity well. The outer electrons in metal atoms readily move from one atom to the next to transmit current.

Figure 15 In metallic bonding, the outer electrons of the silver atoms are not attached to any one silver atom. This allows them to move and conduct electricity.

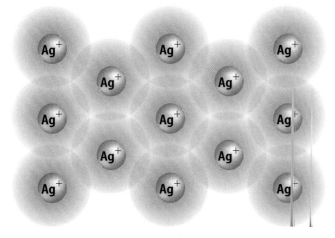

Covalent Bonds—Sharing

Some atoms are unlikely to lose or gain electrons because the number of electrons in their outer levels makes this difficult. For example, carbon has six protons and six electrons. Four of the six electrons are in its outer energy level. To obtain a more stable structure, carbon would either have to gain or lose four electrons. This is difficult because gaining and losing so many electrons takes so much energy. The alternative is sharing electrons.

The Covalent Bond Atoms of many elements become more stable by sharing electrons. The chemical bond that forms between nonmetal atoms when they share electrons is called a **covalent** (koh VAY luhnt) **bond.** Shared electrons are attracted to the nuclei of both atoms. They move back and forth between the outer energy levels of each atom in the covalent bond. So, each atom has a stable outer energy level some of the time. Covalently bonded compounds are called molecular compounds.

✔ **Reading Check** *How do atoms form covalent bonds?*

The atoms in a covalent bond form a neutral particle, which contains the same numbers of positive and negative charges. The neutral particle formed when atoms share electrons is called a **molecule** (MAH lih kyewl). A molecule is the basic unit of a molecular compound. You can see how molecules form by sharing electrons in **Figure 16.** Notice that no ions are involved because no electrons are gained or lost. Crystalline solids, such as sodium chloride, are not referred to as molecules, because their basic units are ions, not molecules.

Mini LAB

Constructing a Model of Methane

Procedure

1. Using **circles of colored paper** to represent protons, neutrons, and electrons, build paper models of one carbon atom and four hydrogen atoms.
2. Use your models of atoms to construct a molecule of methane by forming covalent bonds. The methane molecule has four hydrogen atoms chemically bonded to one carbon atom.

Analysis

1. In the methane molecule, do the carbon and hydrogen atoms have the same arrangement of electrons as two noble gas elements? Explain your answer.
2. Does the methane molecule have a charge?

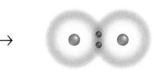

Figure 16 Covalent bonding is another way that atoms become more stable. Sharing electrons allows each atom to have a stable outer energy level. These atoms form a single covalent bond.

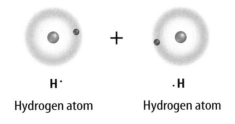

H· + ·H → H:H
Hydrogen atom   Hydrogen atom   Hydrogen molecule

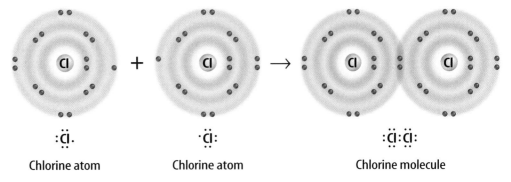

:C̈l· + ·C̈l: → :C̈l:C̈l:
Chlorine atom   Chlorine atom   Chlorine molecule

Figure 17 An atom can also form a covalent bond by sharing two or three electrons.

Carbon atom   Oxygen atoms   Carbon dioxide molecule

In carbon dioxide, carbon shares two electrons with each of two oxygen atoms forming two double bonds. Each oxygen atom shares two electrons with the carbon atom.

Nitrogen atoms   Nitrogen molecule

Each nitrogen atom shares three electrons in forming a triple bond.

Double and Triple Bonds Sometimes an atom shares more than one electron with another atom. In the molecule carbon dioxide, shown in **Figure 17,** each of the oxygen atoms shares two electrons with the carbon atom. The carbon atom shares two of its electrons with each oxygen atom. When two pairs of electrons are involved in a covalent bond, the bond is called a double bond. **Figure 17** also shows the sharing of three pairs of electrons between two nitrogen atoms in the nitrogen molecule. When three pairs of electrons are shared by two atoms, the bond is called a triple bond.

✓ **Reading Check** *How many pairs of electrons are shared in a double bond?*

Polar and Nonpolar Molecules

You have seen how atoms can share electrons and that they become more stable by doing so, but do they always share electrons equally? The answer is no. Some atoms have a greater attraction for electrons than others do. Chlorine, for example, attracts electrons more strongly than hydrogen does. When a covalent bond forms between hydrogen and chlorine, the shared pair of electrons tends to spend more time near the chlorine atom than the hydrogen atom.

This unequal sharing makes one side of the bond more negative than the other, like poles on a battery. This is shown in **Figure 18.** Such bonds are called polar bonds. A **polar bond** is a bond in which electrons are shared unevenly. The bonds between the oxygen atom and hydrogen atoms in the water molecule are another example of polar bonds.

Figure 18 Hydrogen chloride is a polar covalent molecule.

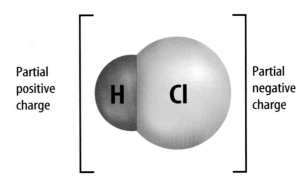

Partial positive charge

Partial negative charge

The Polar Water Molecule Water molecules form when hydrogen and oxygen share electrons. **Figure 19** shows how this sharing is unequal. The oxygen atom has a greater share of the electrons in each bond—the oxygen end of a water molecule has a slight negative charge and the hydrogen end has a slight positive charge. Because of this, water is said to be polar—having two opposite ends or poles like a magnet.

When they are exposed to a negative charge, the water molecules line up like magnets with their positive ends facing the negative charge. You can see how they are drawn to the negative charge on the balloon in **Figure 19.** Water molecules also are attracted to each other. This attraction between water molecules accounts for many of the physical properties of water.

Molecules that do not have these uneven charges are called nonpolar molecules. Because each element differs slightly in its ability to attract electrons, the only completely nonpolar bonds are bonds between atoms of the same element. One example of a nonpolar bond is the triple bond in the nitrogen molecule.

Like ionic compounds, some molecular compounds can form crystals, in which the basic unit is a molecule. Often you can see the pattern of the units in the shape of ionic and molecular crystals, as shown in **Figure 20.**

Science Online

Topic: Polar Molecules
Visit bookl.msscience.com for Web links to information about soaps and detergents.

Activity Oil and water are not soluble in one another. However, if you add a few grams of a liquid dish detergent, the oil will become soluble in the water. Instead of two layers, there will be only one. Explain why soap can help the oil become soluble in water.

Figure 19 Two hydrogen atoms share electrons with one oxygen atom, but the sharing is unequal. The electrons are more likely to be closer to the oxygen than the hydrogens. The space-saving model shows how the charges are separated or polarized. **Define** *the term* polar.

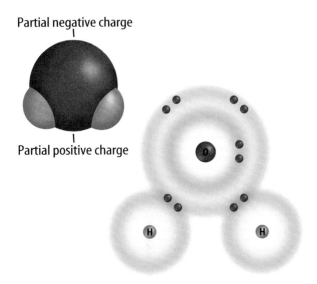

Partial negative charge

Partial positive charge

The positive ends of the water molecules are attracted to the negatively charged balloon, causing the stream of water to bend.

Figure 20

Many solids exist as crystals. Whether tiny grains of table salt or big, chunky blocks of quartz you might find rock hunting, a crystal's shape is often a reflection of the arrangement of its particles. Knowing a solid's crystal structure helps researchers understand its physical properties. Some crystals with cubic and hexagonal shapes are shown here.

Water

O

Si

HEXAGONAL Quartz crystals, above, are six sided, just as a snowflake, above right, has six points. This is because the molecules that make up both quartz and snowflakes arrange themselves into hexagonal patterns.

Ca^{2+}

F^-

Na^+

Cl^-

CUBIC Salt, left, and fluorite, above, form cube-shaped crystals. This shape is a reflection of the cube-shaped arrangement of the ions in the crystal.

Chemical Shorthand

In medieval times, alchemists (AL kuh mists) were the first to explore the world of chemistry. Although many of them believed in magic and mystical transformations, alchemists did learn much about the properties of some elements. They even used symbols to represent them in chemical processes, some of which are shown in **Figure 21.**

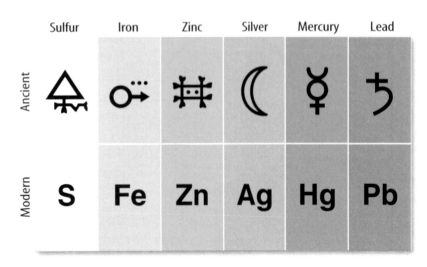

	Sulfur	Iron	Zinc	Silver	Mercury	Lead
Ancient						
Modern	S	Fe	Zn	Ag	Hg	Pb

Symbols for Atoms Modern chemists use symbols to represent elements, too. These symbols can be understood by chemists everywhere. Each element is represented by a one letter-, two letter-, or three-letter symbol. Many symbols are the first letters of the element's name, such as H for hydrogen and C for carbon. Others are the first letters of the element's name in another language, such as K for potassium, which stands for kalium, the Latin word for potassium.

Symbols for Compounds Compounds can be described using element symbols and numbers. For example, **Figure 22** shows how two hydrogen atoms join together in a covalent bond. The resulting hydrogen molecule is represented by the symbol H_2. The small 2 after the H in the formula is called a subscript. *Sub* means "below" and *script* means "write," so a subscript is a number that is written a little below a line of text. The subscript 2 means that two atoms of hydrogen are in the molecule.

Figure 21 Alchemists used elaborate symbols to describe elements and processes. Modern chemical symbols are letters that can be understood all over the world.

Figure 22 Chemical formulas show you the kind and number of atoms in a molecule.
Describe *the term* subscript.

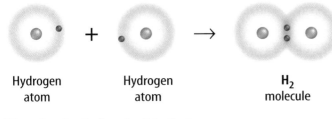

Hydrogen atom + Hydrogen atom → H_2 molecule

The subscript 2 after the H indicates that the hydrogen molecule contains two atoms of hydrogen.

The formula for ammonia, NH_3, tells you that the ratio is one nitrogen atom to three hydrogen atoms.

NH_3

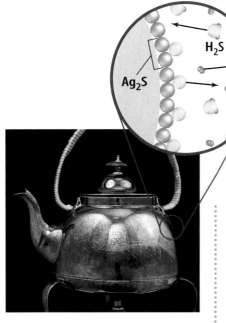

Chemical Formulas A **chemical formula** is a combination of chemical symbols and numbers that shows which elements are present in a compound and how many atoms of each element are present. When no subscript is shown, the number of atoms is understood to be one.

Reading Check *What is a chemical formula and what does it tell you about a compound?*

Now that you understand chemical formulas, you can look back at the other chemical compounds shown earlier in this chapter, and write their chemical formulas. A water molecule contains one oxygen atom and two hydrogen atoms, so its formula is H_2O. Ammonia, shown in **Figure 22,** is a covalent compound that contains one nitrogen atom and three hydrogen atoms. Its chemical formula is NH_3.

The black tarnish that forms on silver, shown in **Figure 23,** is a compound made up of the elements silver and sulfur in the proportion of two atoms of silver to one atom of sulfur. If alchemists knew the composition of silver tarnish, how might they have written a formula for the compound? The modern formula for silver tarnish is Ag_2S. The formula tells you that it is a compound that contains two silver atoms and one sulfur atom.

Figure 23 Silver tarnish is the compound silver sulfide, Ag_2S. The formula shows that two silver atoms are combined with one sulfur atom.

section ② review

Summary

Four Types of Bonds

- Ionic bond is the attraction that holds ions close together.
- Metallic bonds form when metal atoms pool their electrons.
- Covalent bonds form when atoms share electrons.
- A polar covalent bond is a bond in which electrons are shared unevenly.

Chemical Shorthand

- Compounds can be described by using element symbols and numbers.
- A chemical formula is a combination of element symbols and numbers.

Self Check

1. **Determine** Use the periodic table to decide whether lithium forms a positive or negative ion. Does fluorine form a positive or negative ion? Write the formula for the compound formed from these two elements.

2. **Compare and contrast** polar and nonpolar bonds.

3. **Explain** how a chemical formula indicates the ratio of elements in a compound.

4. **Think Critically** Silicon has four electrons in its outer energy level. What type of bond is silicon most likely to form with other elements? Explain.

Applying Skills

5. **Predict** what type of bonds that will form between the following pairs of atoms: carbon and oxygen, potassium and bromine, fluorine and fluorine.

Science Online bookl.msscience.com/self_check_quiz

IONIC COMPOUNDS

Metals in Groups 1 and 2 often lose electrons and form positive ions. Nonmetals in Groups 16 and 17 often gain electrons and become negative ions. How can compounds form between these four groups of elements?

▶ Real-World Question

How do different atoms combine with each other to form compounds?

Goals
- ■ **Construct** models of electron gain and loss.
- ■ **Determine** formulas for the ions and compounds that form when electrons are gained or lost.

Materials
paper (8 different colors)
tacks (2 different colors)
corrugated cardboard
scissors

Safety Precautions

▶ Procedure

1. Cut colored-paper disks 7 cm in diameter to represent the elements Li, S, Mg, O, Ca, Cl, Na, and I. Label each disk with one symbol.

2. Lay circles representing the atoms Li and S side by side on cardboard.

3. Choose colored thumbtacks to represent the outer electrons of each atom. Place the tacks evenly around the disks to represent the outer electron levels of the elements.

4. Move electrons from the metal atom to the nonmetal atom so that both elements

achieve noble gas arrangements of eight outer electrons. If needed, cut additional paper disks to add more atoms of one element.

5. Write the formula for each ion and the compound formed when you shift electrons.

6. Repeat steps 2 through 6 to combine Mg and O, Ca and Cl, and Na and I.

▶ Conclude and Apply

1. **Draw** electron dot diagrams for all of the ions produced.

2. **Identify** the noble gas elements having the same electron arrangements as the ions you made in this lab.

3. **Analyze Results** Why did you have to use more than one atom in some cases? Why couldn't you take more electrons from one metal atom or add extra ones to a nonmetal atom?

*C*ommunicating
Your Data

Compare your compounds and dot diagrams with those of other students in your class. **For more help, refer to the Science Skill Handbook.**

Model and Invent

At🞷mic Structure

Goals

■ **Design** a model of a chosen element.

■ **Observe** the models made by others in the class and identify the elements they represent.

Possible Materials

magnetic board
rubber magnetic strips
candy-coated chocolates
scissors
paper
marker
coins

Safety Precautions

WARNING: *Never eat any food in the laboratory. Wash hands thoroughly.*

⊙ *Real-World Question*

As more information has become known about the structure of the atom, scientists have developed new models. Making your own model and studying the models of others will help you learn how protons, neutrons, and electrons are arranged in an atom. Can an element be identified based on a model that shows the arrangement of the protons, neutrons, and electrons of an atom? How will your group construct a model of an element that others will be able to identify?

⊙ *Make A Model*

1. Choose an element from periods 2 or 3 of the periodic table. How can you determine the number of protons, neutrons, and electrons in an atom given the atom's mass number?

2. How can you show the difference between protons and neutrons? What materials will you use to represent the electrons of the atom? How will you represent the nucleus?

3. How will you model the arrangement of electrons in the atom? Will the atom have a charge? Is it possible to identify an atom by the number of protons it has?

4. Make sure your teacher approves your plan before you proceed.

◉ Test Your Model

1. **Construct** your model. Then record your observations in your Science Journal and include a sketch.

2. **Construct** another model of a different element.

3. **Observe** the models made by your classmates. Identify the elements they represent.

◉ Analyze Your Data

1. **State** what elements you identified using your classmates' models.

2. **Identify** which particles always are present in equal numbers in a neutral atom.

3. **Predict** what would happen to the charge of an atom if one of the electrons were removed.

4. **Describe** what happens to the charge of an atom if two electrons are added. What happens to the charge of an atom if one proton and one electron are removed?

5. **Compare and contrast** your model with the electron cloud model of the atom. How is your model similar? How is it different?

◉ Conclude and Apply

1. **Define** the minimum amount of information that you need to know in order to identify an atom of an element.

2. **Explain** If you made models of the isotopes boron-10 and boron-11, how would these models be different?

Communicating Your Data

Compare your models with those of other students. Discuss any differences you find among the models.

"Baring the Atom's Mother Heart"
from Selu: Seeking the Corn-Mother's Wisdom
by Marilou Awiakta

Author Marilou Awiakta was raised near Oak Ridge National Laboratory, a nuclear research laboratory in Tennessee where her father worked. She is of Cherokee and Irish descent. This essay resulted from conversations the author had with writer Alice Walker. It details the author's concern with nuclear technology.

"What is the atom, Mother? Will it hurt us?"

I was nine years old. It was December 1945. Four months earlier, in the heat of an August morning—Hiroshima. Destruction. Death. Power beyond belief, released from something invisible[1]. Without knowing its name, I'd already felt the atoms' power in another form...

"What is the atom, Mother? Will it hurt us?"

"It can be used to hurt everybody, Marilou. It killed thousands[2] of people in Hiroshima and Nagasaki. But the atom itself. . . ? It's invisible, the smallest bit of matter. And it's in everything. Your hand, my dress, the milk you're drinking—. . .

. . . Mother already had taught me that beyond surface differences, everything is [connected]. It seemed natural for the atom to be part of this connection. At school, when I was introduced to Einstein's theory of relativity—that energy and matter are one—I accepted the concept easily.

1 can't see

2 10,500

Understanding Literature

Refrain Refrains are emotionally charged words or phrases that are repeated throughout a literary work and can serve a number of purposes. In this work, the refrain is when the author asks, "What is the atom, Mother? Will it hurt us?" Do you think the refrain helps the reader understand the importance of the atom?

Respond to the Reading

1. How did the author's mother explain the atom to her?
2. Is this a positive or negative explanation of the atom?
3. **Linking Science and Writing** Write a short poem about some element you learned about in this chapter.

Nuclear fission, or splitting atoms, is the breakdown of an atom's nucleus. It occurs when a particle, such as a neutron, strikes the nucleus of a uranium atom, splitting the nucleus into two fragments, called fission fragments, and releasing two or three neutrons. These released neutrons ultimately cause a chain reaction by splitting more nuclei and releasing more neutrons. When it is uncontrolled, this chain reaction results in a devastating explosion.

Reviewing Main Ideas

Section 1 Why do atoms combine?

1. The electrons in the electron cloud of an atom are arranged in energy levels.

2. Each energy level can hold a specific number of electrons.

3. The periodic table supplies a great deal of information about the elements.

4. The number of electrons in an atom increases across each period of the periodic table.

5. The noble gas elements are stable because their outer energy levels are stable.

6. Electron dot diagrams show the electrons in the outer energy level of an atom.

Section 2 How Elements Bond

1. An atom can become stable by gaining, losing, or sharing electrons so that its outer energy level is full.

2. Ionic bonds form when a metal atom loses one or more electrons and a nonmetal atom gains one or more electrons.

3. Covalent bonds are created when two or more nonmetal atoms share electrons.

4. The unequal sharing of electrons results in a polar covalent bond.

5. A chemical formula indicates the kind and number of atoms in a compound.

Visualizing Main Ideas

Copy and complete the following concept map on types of bonds.

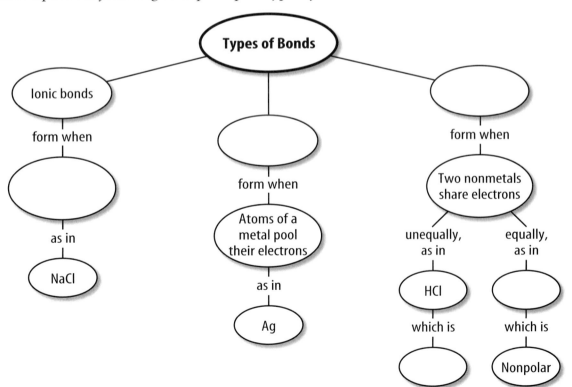

Using Vocabulary

chemical bond p. 15
chemical formula p. 24
compound p. 17
covalent bond p. 19
electron cloud p. 8
electron dot diagram
   p. 14

energy level p. 9
ion p. 17
ionic bond p. 17
metallic bond p. 18
molecule p. 19
polar bond p. 20

Distinguish between the terms in each of the following pairs.

1. ion—molecule

2. molecule—compound

3. electron dot diagram—ion

4. chemical formula—molecule

5. ionic bond—covalent bond

6. electron cloud—electron dot diagram

7. covalent bond—polar bond

8. compound—formula

9. metallic bond—ionic bond

Checking Concepts

Choose the word or phrase that best answers the question.

10. Which of the following is a covalently bonded molecule?
    A) Cl_2
    B) air
    C) Ne
    D) salt

11. What is the number of the group in which the elements have a stable outer energy level?
    A) 1
    B) 13
    C) 16
    D) 18

12. Which term describes the units that make up substances formed by ionic bonding?
    A) ions
    B) molecules
    C) acids
    D) atoms

13. Which of the following describes what is represented by the symbol Cl^-?
    A) an ionic compound
    B) a polar molecule
    C) a negative ion
    D) a positive ion

14. What happens to electrons in the formation of a polar covalent bond?
    A) They are lost.
    B) They are gained.
    C) They are shared equally.
    D) They are shared unequally.

15. Which of the following compounds is unlikely to contain ionic bonds?
    A) NaF
    B) CO
    C) LiCl
    D) $MgBr_2$

16. Which term describes the units that make up compounds with covalent bonds?
    A) ions
    B) molecules
    C) salts
    D) acids

17. In the chemical formula CO_2, the subscript 2 shows which of the following?
    A) There are two oxygen ions.
    B) There are two oxygen atoms.
    C) There are two CO_2 molecules.
    D) There are two CO_2 compounds.

Use the figure below to answer question 18.

18. Which is NOT true about the molecule H_2O?
    A) It contains two hydrogen atoms.
    B) It contains one oxygen atom.
    C) It is a polar covalent compound.
    D) It is an ionic compound.

Science Online bookl.msscience.com/vocabulary_puzzlemaker

Thinking Critically

19. **Explain** why Groups 1 and 2 form many compounds with Groups 16 and 17.

Use the illustration below to answer questions 20 and 21.

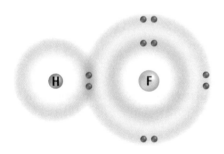

20. **Explain** what type of bond is shown here.

21. **Predict** In the HF molecule above, predict if the electrons are shared equally or unequally between the two atoms. Where do the electrons spend more of their time?

22. **Analyze** When salt dissolves in water, the sodium and chloride ions separate. Explain why this might occur.

23. **Interpret Data** Both cesium, in period 6, and lithium, in period 2, are in the alkali metals family. Cesium is more reactive. Explain this using the energy step diagram in **Figure 4**.

24. **Explain** Use the fact that water is a polar molecule to explain why water has a much higher boiling point than other molecules of its size.

25. **Predict** If equal masses of CuCl and $CuCl_2$ decompose into their components— copper and chlorine—predict which compound will yield more copper. Explain.

26. **Concept Map** Draw a concept map starting with the term *Chemical Bond* and use all the vocabulary words.

27. **Recognize Cause and Effect** A helium atom has only two electrons. Why does helium behave as a noble gas?

28. **Draw a Conclusion** A sample of an element can be drawn easily into wire and conducts electricity well. What kind of bonds can you conclude are present?

Performance Activities

29. **Display** Make a display featuring one of the element families described in this chapter. Include electronic structures, electron dot diagrams, and some compounds they form.

Applying Math

Use the table below to answer question 30.

Formulas of Compounds

Compound	Number of Metal Atoms	Number of Nonmetal Atoms
Cu_2O		
Al_2S_3	Do not write in this book.	
NaF		
$PbCl_4$		

30. **Make and Use Tables** Fill in the second column of the table with the number of metal atoms in one unit of the compound. Fill in the third column with the number of atoms of the nonmetal in one unit.

31. **Molecules** What are the percentages of each atom for this molecule, K_2CO_3?

32. **Ionic Compounds** Lithium, as a positive ion, is written as Li^{1+}. Nitrogen, as a negative ion, is written as N^{3-}. In order for the molecule to be neutral, the plus and minus charges have to equal zero. How many lithium atoms are needed to make the charges equal to zero?

33. **Energy Levels** Calculate the maximum number of electrons in energy level 6.

Part 1 **Multiple Choice**

Record your answers on the answer sheet provided by your teacher or on a sheet of paper.

1. Sodium combines with fluorine to produce sodium fluoride (NaF), an active ingredient in toothpaste. In this form, sodium has the electron configuration of which other element?
   A. neon
   B. magnesium
   C. lithium
   D. chlorine

Use the illustration below to answer questions 2 and 3.

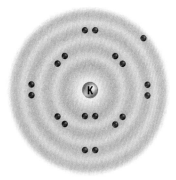

2. The illustration above shows the electron configuration for potassium. How many electrons does potassium need to gain or lose to become stable?
   A. gain 1
   B. gain 2
   C. lose 1
   D. lose 2

3. Potassium belongs to the Group 1 family of elements on the periodic table. What is the name of this group?
   A. halogens
   B. alkali metals
   C. noble gases
   D. alkaline metals

4. What type of bond connects the atoms in a molecule of nitrogen gas (N_2)?
   A. ionic
   B. single
   C. double
   D. triple

Use the illustration below to answer questions 5 and 6.

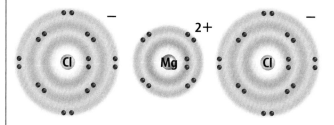

Magnesium chloride

5. The illustration above shows the electron distribution for magnesium chloride. Which of the following is the correct way to write the formula for magnesium chloride?
   A. Mg_2Cl
   B. $MgCl_2$
   C. $MgCl$
   D. Mg_2Cl_2

6. Which of the following terms best describes the type of bonding in magnesium chloride?
   A. ionic
   B. pooling
   C. metallic
   D. covalent

7. What is the maximum number of electrons in the third energy level?
   A. 8
   B. 16
   C. 18
   D. 24

Part 2 | Short Response/Grid In

Record your answers on the answer sheet provided by your teacher or on a sheet of paper.

8. What is an electron cloud?

9. Explain what is wrong with the following statement: All covalent bonds between atoms are polar to some degree because each element differs slightly in its ability to attract electrons. Give an example to support your answer.

Use the illustration below to answer questions 10 and 11.

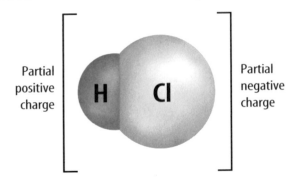

Partial positive charge — H Cl — Partial negative charge

10. The illustration above shows how hydrogen and chlorine combine to form a polar molecule. Explain why the bond is polar.

11. What is the electron dot diagram for the molecule in the illustration?

12. What is the name of the family of elements in Group 17 of the periodic table?

13. Name two ways that electrons around a nucleus are different from planets circling the Sun.

14. Which family of elements used to be known as inert gases? Why was the name changed?

Test-Taking Tip

Take Your Time Stay focused during the test and don't rush, even if you notice that other students are finishing the test early.

Part 3 | Open Ended

Record your answers on a sheet of paper.

15. Scientific experiments frequently require an oxygen-free environment. Such experiments often are performed in containers flooded with argon gas. Describe the arrangement of electrons in an argon atom. Why is argon often a good choice for these experiments?

16. Which group of elements is called the halogen elements? Describe their electron configurations and discuss their reactivity. Name two elements that belong to this group.

17. What is an ionic bond? Describe how sodium chloride forms an ionic bond.

18. Explain metallic bonding. What are some ways this affects the properties of metals?

19. Explain why polar molecules exist, but polar ionic compounds do not exist.

Use the illustration below to answer questions 20 and 21.

20. Explain what is happening in the photograph above. What would happen if the balloon briefly touched the water?

21. Draw a model showing the electron distribution for a water molecule. Explain how the position of the electrons causes the effect shown in your illustration.

Chemical Reactions

What chemical reactions happen at chemical plants?

Chemical plants like this one provide the starting materials for thousands of chemical reactions. The compact discs you listen to, personal items, such as shampoo and body lotion, and medicines all have their beginnings in a chemical plant.

Science Journal What additional types of products do you think are manufactured in a chemical plant?

Start-Up Activities

Identify a Chemical Reaction

You can see substances changing every day. Fuels burn, giving energy to cars and trucks. Green plants convert carbon dioxide and water into oxygen and sugar. Cooking an egg or baking bread causes changes too. These changes are called chemical reactions. In this lab you will observe a common chemical change.

WARNING: *Do not touch the test tube. It will be hot. Use extreme caution around an open flame. Point test tubes away from you and others.*

1. Place 3 g of sugar into a large test tube.

2. Carefully light a laboratory burner.

3. Using a test-tube holder, hold the bottom of the test tube just above the flame for 45 s or until something happens with the sugar.

4. Observe any change that occurs.

5. **Think Critically** Describe in your Science Journal the changes that took place in the test tube. What do you think happened to the sugar? Was the substance that remained in the test tube after heating the same as the substance you started with?

Chemical Reaction Make the following Foldable to help you understand chemical reactions.

STEP 1 Fold a vertical sheet of notebook paper in half lengthwise.

STEP 2 Cut along every third line of only the top layer to form tabs.

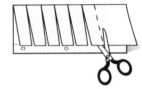

STEP 3 Label each tab.

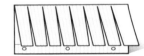

Research Information Before you read the chapter, write several questions you have about chemical reactions on the front of the tabs. As you read, add more questions. Under the tabs of your Foldable, write answers to the questions you recorded on the tabs.

Preview this chapter's content and activities at
bookl.msscience.com

Chemical Formulas and Equations

as you read

What You'll Learn

- **Determine** whether or not a chemical reaction is occurring.
- **Determine** how to read and understand a balanced chemical equation.
- **Examine** some reactions that release energy and others that absorb energy.
- **Explain** the law of conservation of mass.

Why It's Important

Chemical reactions warm your home, cook your meals, digest your food, and power cars and trucks.

Review Vocabulary

atom: the smallest piece of matter that still retains the property of the element

New Vocabulary

- chemical reaction
- reactant
- product
- chemical equation
- endothermic reaction
- exothermic reaction

Physical or Chemical Change?

You can smell a rotten egg and see the smoke from a campfire. Signs like these tell you that a chemical reaction is taking place. Other evidence might be less obvious, but clues are always present to announce that a reaction is under way.

Matter can undergo two kinds of changes—physical and chemical. Physical changes in a substance affect only physical properties, such as its size and shape, or whether it is a solid, liquid, or gas. For example, when water freezes, its physical state changes from liquid to solid, but it's still water.

In contrast, chemical changes produce new substances that have properties different from those of the original substances. The rust on a bike's handlebars, for example, has properties different from those of the metal around it. Another example is the combination of two liquids that produce a precipitate, which is a solid, and a liquid. The reaction of silver nitrate and sodium chloride forms solid silver chloride and liquid sodium nitrate. A process that produces chemical change is a **chemical reaction.**

To compare physical and chemical changes, look at the newspaper shown in **Figure 1.** If you fold it, you change its size and shape, but it is still newspaper. Folding is a physical change. If you use it to start a fire, it will burn. Burning is a chemical change because new substances result. How can you recognize a chemical change? **Figure 2** shows what to look for.

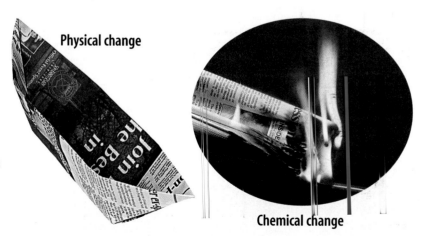

Physical change

Chemical change

Figure 1 Newspaper can undergo both physical and chemical changes.

Figure 2

Chemical reactions take place when chemicals combine to form new substances. Your senses—sight, taste, hearing, smell, and touch—can help you detect chemical reactions in your environment.

▼ **TASTE** A boy grimaces after sipping milk that has gone sour due to a chemical reaction.

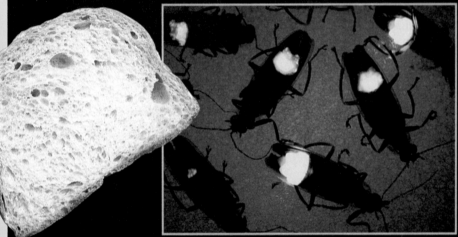

▲ **SIGHT** When you spot a firefly's bright glow, you are seeing a chemical reaction in progress—two chemicals are combining in the firefly's abdomen and releasing light in the process. The holes in a slice of bread are visible clues that sugar molecules were broken down by yeast cells in a chemical reaction that produces carbon dioxide gas. The gas caused the bread dough to rise.

▲ **SMELL AND TOUCH** Billowing clouds of acrid smoke and waves of intense heat indicate that chemical reactions are taking place in this burning forest.

▲ **HEARING** A Russian cosmonaut hoists a flare into the air after landing in the ocean during a training exercise. The hissing sound of the burning flare is the result of a chemical reaction.

Chemical Equations

To describe a chemical reaction, you must know which substances react and which substances are formed in the reaction. The substances that react are called the reactants (ree AK tunts). **Reactants** are the substances that exist before the reaction begins. The substances that form as a result of the reaction are called the **products.**

When you mix baking soda and vinegar, a vigorous chemical reaction occurs. The mixture bubbles and foams up inside the container, as you can see in **Figure 3.**

Baking soda and vinegar are the common names for the reactants in this reaction, but they also have chemical names. Baking soda is the compound sodium hydrogen carbonate (often called sodium bicarbonate), and vinegar is a solution of acetic (uh SEE tihk) acid in water. What are the products? You saw bubbles form when the reaction occurred, but is that enough of a description?

Describing What Happens Bubbles tell you that a gas has been produced, but they don't tell you what kind of gas. Are bubbles of gas the only product, or do some atoms from the vinegar and baking soda form something else? What goes on in the chemical reaction can be more than what you see with your eyes. Chemists try to find out which reactants are used and which products are formed in a chemical reaction. Then, they can write it in a shorthand form called a chemical equation. A **chemical equation** tells chemists at a glance the reactants, products, physical state, and the proportions of each substance present. This is very important as you will see later.

Reading Check *What does a chemical equation tell chemists?*

Figure 3 The bubbles tell you that a chemical reaction has taken place.
Predict *how you might find out whether a new substance has formed.*

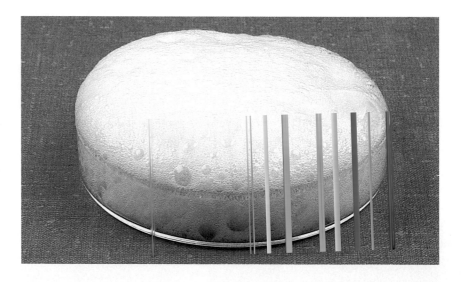

Table 1 Reactions Around the Home

Reactants		Products
Baking soda + Vinegar	→	Gas + White solid
Charcoal + Oxygen	→	Ash + Gas +Heat
Iron + Oxygen + Water	→	Rust
Silver + Hydrogen sulfide	→	Black tarnish + Gas
Gas (kitchen range) + Oxygen	→	Gas + Heat
Sliced apple + Oxygen	→	Apple turns brown

Using Words One way you can describe a chemical reaction is with an equation that uses words to name the reactants and products. The reactants are listed on the left side of an arrow, separated from each other by plus signs. The products are placed on the right side of the arrow, also separated by plus signs. The arrow between the reactants and products represents the changes that occur during the chemical reaction. When reading the equation, the arrow is read as *produces.*

You can begin to think of processes as chemical reactions even if you do not know the names of all the substances involved. **Table 1** can help you begin to think like a chemist. It shows the word equations for chemical reactions you might see around your home. See how many other reactions you can find. Look for the signs you have learned that indicate a reaction might be taking place. Then, try to write them in the form shown in the table.

Using Chemical Names Many chemicals used around the home have common names. For example, acetic acid dissolved in water is called vinegar. Some chemicals, such as baking soda, have two common names—it also is known as sodium bicarbonate. However, chemical names are usually used in word equations instead of common names. In the baking soda and vinegar reaction, you already know the chemical names of the reactants—sodium hydrogen carbonate and acetic acid. The names of the products are sodium acetate, water, and carbon dioxide. The word equation for the reaction is as follows.

Acetic acid + Sodium hydrogen carbonate →
Sodium acetate + Water + Carbon dioxide

Autumn Leaves A color change can indicate a chemical reaction. When leaves change colors in autumn, the reaction may not be what you expect. The bright yellow and orange are always in the leaves, but masked by green chlorophyll. When the growth season ends, more chlorophyll is broken down than produced. The orange and yellow colors become visible.

Observing the Law of Conservation of Mass

Procedure

1. Place a piece of **steel wool** into a **medium test tube**. Seal the end of the test tube with a **balloon**.
2. Find the mass.
3. Using a test-tube holder, heat the bottom of the tube for two minutes in a **hot water bath** provided by your teacher. Allow the tube to cool completely.
4. Find the mass again.

Analysis

1. What did you observe that showed a chemical reaction took place?
2. Compare the mass before and after the reaction.
3. Why was it important for the test tube to be sealed?

Using Formulas The word equation for the reaction of baking soda and vinegar is long. That's why chemists use chemical formulas to represent the chemical names of substances in the equation. You can convert a word equation into a chemical equation by substituting chemical formulas for the chemical names. For example, the chemical equation for the reaction between baking soda and vinegar can be written as follows:

$$CH_3COOH + NaHCO_3 \rightarrow CH_3COONa + H_2O + CO_2$$

| Acetic acid (vinegar) | Sodium hydrogen carbonate (baking soda) | Sodium acetate | Water | Carbon dioxide |

Subscripts When you look at chemical formulas, notice the small numbers written to the right of the atoms. These numbers, called subscripts, tell you the number of atoms of each element in that compound. For example, the subscript 2 in CO_2 means that each molecule of carbon dioxide has two oxygen atoms. If an atom has no subscript, it means that only one atom of that element is in the compound, so carbon dioxide has only one carbon atom.

Conservation of Mass

What happens to the atoms in the reactants when they are converted into products? According to the law of conservation of mass, the mass of the products must be the same as the mass of the reactants in that chemical reaction. This principle was first stated by the French chemist Antoine Lavoisier (1743–1794), who is considered the first modern chemist. Lavoisier used logic and scientific methods to study chemical reactions. He proved by his experiments that nothing is lost or created in chemical reactions.

He showed that chemical reactions are much like mathematical equations. In math equations, the right and left sides of the equation are numerically equal. Chemical equations are similar, but it is the number and kind of atoms that are equal on the two sides. Every atom that appears on the reactant side of the equation also appears on the product side, as shown in **Figure 4.** Atoms are never lost or created in a chemical reaction; however, they do change partners.

Figure 4 The law of conservation of mass states that the number and kind of atoms must be equal for products and reactants.

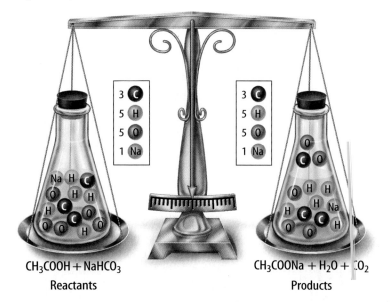

CH$_3$COOH + NaHCO$_3$
Reactants

CH$_3$COONa + H$_2$O + CO$_2$
Products

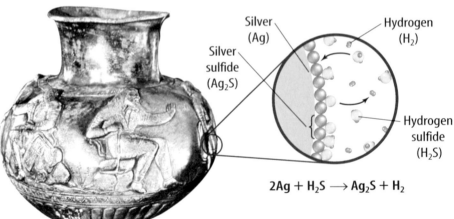

Figure 5 Keeping silver bright takes frequent polishing, especially in homes heated by gas. Sulfur compounds found in small concentrations in natural gas react with silver, forming black silver sulfide, Ag_2S.

Silver (Ag)

Silver sulfide (Ag_2S)

Hydrogen (H_2)

Hydrogen sulfide (H_2S)

$$2Ag + H_2S \longrightarrow Ag_2S + H_2$$

Balancing Chemical Equations

When you write the chemical equation for a reaction, you must observe the law of conservation of mass. Look back at **Figure 4.** It shows that when you count the number of carbon, hydrogen, oxygen, and sodium atoms on each side of the arrow in the equation, you find equal numbers of each kind of atom. This means the equation is balanced and the law of conservation of mass is observed.

Not all chemical equations are balanced so easily. For example, silver tarnishes, as in **Figure 5,** when it reacts with sulfur compounds in the air, such as hydrogen sulfide. The following unbalanced equation shows what happens when silver tarnishes.

$$Ag \quad + \quad H_2S \quad \rightarrow \quad Ag_2S \quad + \quad H_2$$

Silver   Hydrogen sulfide   Silver sulfide   Hydrogen

Count the Atoms Count the number of atoms of each type in the reactants and in the products. The same numbers of hydrogen and sulfur atoms are on each side, but one silver atom is on the reactant side and two silver atoms are on the product side. This cannot be true. A chemical reaction cannot create a silver atom, so this equation does not represent the reaction correctly. Place a 2 in front of the reactant Ag and check to see if the equation is balanced. Recount the number of atoms of each type.

$$2Ag + H_2S \rightarrow Ag_2S + H_2$$

The equation is now balanced. There are an equal number of silver atoms in the reactants and the products. When balancing chemical equations, numbers are placed before the formulas as you did for Ag. These are called coefficients. However, never change the subscripts written to the right of the atoms in a formula. Changing these numbers changes the identity of the compound.

Science nline

Topic: Chemical Equations
Visit bookl.msscience.com for Web links to information about chemical equations and balancing them.

Activity Find a chemical reaction that takes place around your home or school. Write a chemical equation describing it.

Energy in Chemical Reactions

Often, energy is released or absorbed during a chemical reaction. The energy for the welding torch in **Figure 6** is released when hydrogen and oxygen combine to form water.

$$2H_2 + O_2 \rightarrow 2H_2O + \text{energy}$$

Energy Released Where does this energy come from? To answer this question, think about the chemical bonds that break and form when atoms gain, lose, or share electrons. When such a reaction takes place, bonds break in the reactants and new bonds form in the products. In reactions that release energy, the products are more stable, and their bonds have less energy than those of the reactants. The extra energy is released in various forms—light, sound, and heat.

Applying Math   Balancing Equations

CONSERVING MASS Methane and oxygen react to form carbon dioxide, water, and heat. You can see how mass is conserved by balancing the equation: $CH_4 + O_2 \rightarrow CO_2 + H_2O$.

Solution

1 *This is what you know:*

The number of atoms of C, H, and O in reactants and products.

2 *This is what you need to do:*

Make sure that the reactants and products have equal numbers of atoms of each element. Start with the reactant having the greatest number of atoms.

Reactants	Products	Action
$CH_4 + O_2$ have 4 H atoms	$CO_2 + H_2O$ have 2 H atoms	Need 2 more H atoms in Products Multiply H_2O by 2 to give 4 H atoms
$CH_4 + O_2$ have 2 O atoms	$CO_2 + H_2O$ have 4 O atoms	Need 2 more O atoms in Reactants Multiply O_2 by 2 to give 4 O atoms

The balanced equation is $CH_4 + 2O_2 \rightarrow CO_2 + 2H_2O$.

3 *Check your answer:*

Count the carbons, hydrogens, and oxygens on each side.

Practice Problems

1. Balance the equation $Fe_2O_3 + CO \rightarrow Fe_3O_4 + CO_2$.

2. Balance the equation $Al + I_2 \rightarrow AlI_3$.

Science Online   For more practice, visit bookl.msscience.com/ math_practice

Figure 6 This welding torch burns hydrogen and oxygen to produce temperatures above 3,000°C. It can even be used underwater.
Identify *the products of this chemical reaction.*

Energy Absorbed What happens when the reverse situation occurs? In reactions that absorb energy, the reactants are more stable, and their bonds have less energy than those of the products.

$$2H_2O \quad + \quad energy \quad \rightarrow \quad 2H_2 \quad + \quad O_2$$
Water   Hydrogen   Oxygen

In this reaction the extra energy needed to form the products can be supplied in the form of electricity, as shown in **Figure 7.**

As you have seen, reactions can release or absorb energy of several kinds, including electricity, light, sound, and heat. When heat energy is gained or lost in reactions, special terms are used. **Endothermic** (en doh THUR mihk) **reactions** absorb heat energy. **Exothermic** (ek soh THUR mihk) **reactions** release heat energy. You may notice that the root word *therm* refers to heat, as it does in thermos bottles and thermometers.

Heat Released You might already be familiar with several types of reactions that release heat. Burning is an exothermic chemical reaction in which a substance combines with oxygen to produce heat along with light, carbon dioxide, and water.

 *What type of chemical reaction is burning?*

Rapid Release Sometimes energy is released rapidly. For example, charcoal lighter fluid combines with oxygen in the air and produces enough heat to ignite a charcoal fire within a few minutes.

Figure 7 Electrical energy is needed to break water into its components. This is the reverse of the reaction that takes place in the welding torch shown in **Figure 6.**

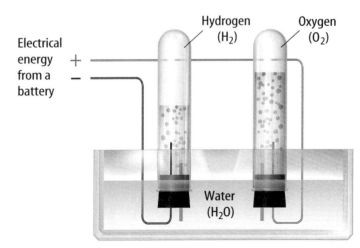

Fast reaction

Figure 8 Two exothermic reactions are shown. The charcoal fire to cook the food was started when lighter fluid combined rapidly with oxygen in air. The iron in the wheelbarrow combined slowly with oxygen in the air to form rust.

Slow reaction

Slow Release Other materials also combine with oxygen but release heat so slowly that you cannot see or feel it happen. This is the case when iron combines with oxygen in the air to form rust. The slow heat release from a reaction also is used in heat packs that can keep your hands warm for several hours. Fast and slow energy release are compared in **Figure 8.**

Heat Absorbed Some chemical reactions and physical processes need to have heat energy added before they can proceed. An example of an endothermic physical process that absorbs heat energy is the cold pack shown in **Figure 9.**

The heavy plastic cold pack holds ammonium nitrate and water. The two substances are separated by a plastic divider. When you squeeze the bag, you break the divider so that the ammonium nitrate dissolves in the water. The dissolving process absorbs heat energy, which must come from the surrounding environment—the surrounding air or your skin after you place the pack on the injury.

Figure 9 The heat energy needed to dissolve the ammonium nitrate in this cold pack comes from the surrounding environment.

Energy in the Equation The word *energy* often is written in equations as either a reactant or a product. Energy written as a reactant helps you think of energy as a necessary ingredient for the reaction to take place. For example, electrical energy is needed to break up water into hydrogen and oxygen. It is important to know that energy must be added to make this reaction occur.

Similarly, in the equation for an exothermic reaction, the word *energy* often is written along with the products. This tells you that energy is released. You include energy when writing the reaction that takes place between oxygen and methane in natural gas when you cook on a gas range, as shown in **Figure 10.** This heat energy cooks your food.

$$CH_4 \;+\; 2O_2 \;\rightarrow\; CO_2 \;+\; 2H_2O \;+\; \text{energy}$$
Methane Oxygen   Carbon   Water
                 dioxide

Although it is not necessary, writing the word *energy* can draw attention to an important aspect of the equation.

Figure 10 Energy from a chemical reaction is used to cook. **Determine** *if energy is used as a reactant or a product in this reaction.*

section 1 review

Summary

Physical or Chemical Change?

- Matter can undergo physical and chemical changes.
- A chemical reaction produces chemical changes.

Chemical Equations

- A chemical equation describes a chemical reaction.
- Chemical formulas represent chemical names for substances.
- A balanced chemical equation has the same number of atoms of each kind on both sides of the equation.

Energy in Chemical Reactions

- Endothermic reactions absorb heat energy.
- Exothermic reactions release heat energy.

Self Check

1. **Determine** if each of these equations is balanced. Why or why not?
   a. $Ca + Cl_2 \rightarrow CaCl_2$
   b. $Zn + Ag_2S \rightarrow ZnS + Ag$

2. **Describe** what evidence might tell you that a chemical reaction has occurred.

3. **Think Critically** After a fire, the ashes have less mass and take up less space than the trees and vegetation before the fire. How can this be explained in terms of the Law of Conservation of Mass?

Applying Math

4. **Calculate** The equation for the decomposition of silver oxide is $2Ag_2O \rightarrow 4Ag + O_2$. Set up a ratio to calculate the number of oxygen molecules released when 1 g of silver oxide is broken down. There are 2.6×10^{21} molecules in 1 g of silver oxide.

Rates of Chemical Reactions

as you read

What You'll Learn

- **Determine** how to describe and measure the speed of a chemical reaction.
- **Identify** how chemical reactions can be speeded up or slowed down.

Why It's Important

Speeding up useful reactions and slowing down destructive ones can be helpful.

⚙ Review Vocabulary

state of matter: physical property that is dependent on temperature and pressure and occurs in four forms—solid, liquid, gas, or plasma

New Vocabulary

- activation energy
- rate of reaction
- concentration
- inhibitor
- catalyst
- enzyme

How Fast?

Fireworks explode in rapid succession on a summer night. Old copper pennies darken slowly while they lie forgotten in a drawer. Cooking an egg for two minutes instead of five minutes makes a difference in the firmness of the yolk. The amount of time you leave coloring solution on your hair must be timed accurately to give the color you want. Chemical reactions are common in your life. However, notice from these examples that time has something to do with many of them. As you can see in **Figure 11,** not all chemical reactions take place at the same rate.

Some reactions, such as fireworks or lighting a campfire, need help to get going. You may also notice that others seem to start on their own. In this section, you will also learn about factors that make reactions speed up or slow down once they get going.

Figure 11 Reaction speeds vary greatly. Fireworks are over in a few seconds. However, the copper coating on pennies darkens slowly as it reacts with substances it touches.

Activation Energy—Starting a Reaction

Before a reaction can start, molecules of the reactants have to bump into each other, or collide. This makes sense because to form new chemical bonds, atoms have to be close together. But, not just any collision will do. The collision must be strong enough. This means the reactants must smash into each other with a certain amount of energy. Anything less, and the reaction will not occur. Why is this true?

To form new bonds in the product, old bonds must break in the reactants, and breaking bonds takes energy. To start any chemical reaction, a minimum amount of energy is needed. This energy is called the **activation energy** of the reaction.

Reading Check *What term describes the minimum amount of energy needed to start a reaction?*

What about reactions that release energy? Is there an activation energy for these reactions too? Yes, even though they release energy later, these reactions also need enough energy to start.

One example of a reaction that needs energy to start is the burning of gasoline. You have probably seen movies in which a car plunges over a cliff, lands on the rocks below, and suddenly bursts into flames. But if some gasoline is spilled accidentally while filling a gas tank, it probably will evaporate harmlessly in a short time.

Why doesn't this spilled gasoline explode as it does in the movies? The reason is that gasoline needs energy to start burning. That is why there are signs at filling stations warning you not to smoke. Other signs advise you to turn off the ignition, not to use mobile phones, and not to reenter the car until fueling is complete.

This is similar to the lighting of the Olympic Cauldron, as shown in **Figure 12.** Cauldrons designed for each Olympics contain highly flammable materials that cannot be extinguished by high winds or rain. However, they do not ignite until the opening ceremonies when a runner lights the cauldron using a flame that was kindled in Olympia, Greece, the site of the original Olympic Games.

Science Online

Topic: Olympic Torch

Visit bookl.msscience.com for Web links to information about the Olympic Torch.

Activity With each new Olympics, the host city devises a new Olympic Torch. Research the process that goes into developing the torch and the fuel it uses.

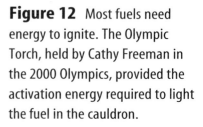

Figure 12 Most fuels need energy to ignite. The Olympic Torch, held by Cathy Freeman in the 2000 Olympics, provided the activation energy required to light the fuel in the cauldron.

Figure 13 The diminishing amount of wax in this candle as it burns indicates the rate of the reaction.

Reaction Rate

Many physical processes are measured in terms of a rate. A rate tells you how much something changes over a given period of time. For example, the rate or speed at which you run or ride your bike is the distance you move divided by the time it took you to move that distance. You may jog at a rate of 8 km/h.

Chemical reactions have rates, too. The **rate of reaction** tells how fast a reaction occurs after it has started. To find the rate of a reaction, you can measure either how quickly one of the reactants is consumed or how quickly one of the products is created, as in **Figure 13.** Both measurements tell how the amount of a substance changes per unit of time.

✔ **Reading Check** *What can you measure to determine the rate of a reaction?*

Reaction rate is important in industry because the faster the product can be made, the less it usually costs. However, sometimes fast rates of reaction are undesirable such as the rates of reactions that cause food spoilage. In this case, the slower the reaction rate, the longer the food will stay edible. What conditions control the reaction rate, and how can the rate be changed?

Temperature Changes Rate You can keep the food you buy at the store from spoiling so quickly by putting it in the refrigerator or freezer, as in **Figure 14.** Food spoiling is a chemical reaction. Lowering the temperature of the food slows the rate of this reaction.

Figure 14 Refrigerated foods must be kept below a certain temperature to slow spoilage. These grapes prove that spoilage, a chemical reaction, has occurred.

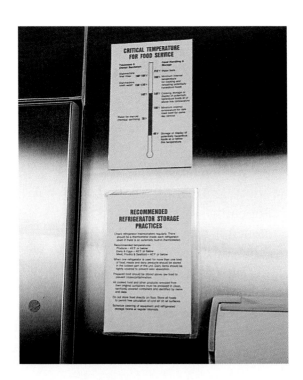

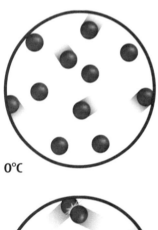

INTEGRATE Health Meat and fish decompose faster at higher temperatures, producing toxins that can make you sick. Keeping these foods chilled slows the decomposition process. Bacteria grow faster at higher temperatures, too, so they reach dangerous levels sooner. Eggs may contain such bacteria, but the heat required to cook eggs also kills bacteria, so hard-cooked eggs are safer to eat than soft-cooked or raw eggs.

Temperature Affects Rate Most chemical reactions speed up when temperature increases. This is because atoms and molecules are always in motion, and they move faster at higher temperatures, as shown in **Figure 15.** Faster molecules collide with each other more often and with greater energy than slower molecules do, so collisions are more likely to provide enough energy to break the old bonds. This is the activation energy.

The high temperature inside an oven speeds up the chemical reactions that turn a liquid cake batter into a more solid, spongy cake. This works the other way, too. Lowering the temperature slows down most reactions. If you set the oven temperature too low, your cake will not bake properly.

Concentration Affects Rate The closer reactant atoms and molecules are to each other, the greater the chance of collisions between them and the faster the reaction rate. It's like the situation shown in **Figure 16.** When you try to walk through a crowded train station, you're more likely to bump into other people than if the station were not so crowded. The amount of substance present in a certain volume is called the **concentration** of that substance. If you increase the concentration, you increase the number of particles of a substance per unit of volume.

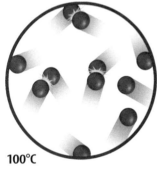

Figure 15 Molecules collide more frequently at higher temperatures than at lower temperatures. This means they are more likely to react.

Collisions are more frequent in a concentrated solution.

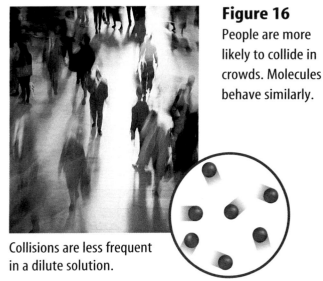

Collisions are less frequent in a dilute solution.

Figure 16 People are more likely to collide in crowds. Molecules behave similarly.

Figure 17 Iron atoms trapped inside this steel beam cannot react with oxygen quickly. More iron atoms are exposed to oxygen molecules in this steel wool, so the reaction speeds up.

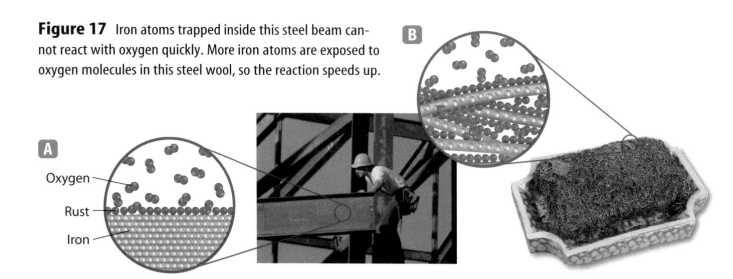

Oxygen

Rust

Iron

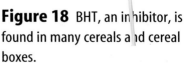

Identifying Inhibitors

Procedure

1. Look at the ingredients listed on **packages of cereals** and **crackers** in your kitchen.
2. Note the preservatives listed. These are chemical inhibitors.
3. Compare the date on the box with the approximate date the box was purchased to estimate shelf life.

Analysis

1. What is the average shelf life of these products?
2. Why is increased shelf life of such products important?

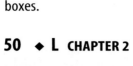

Surface Area Affects Rate The exposed surface area of reactant particles also affects how fast the reaction can occur. You can quickly start a campfire with small twigs, but starting a fire with only large logs would probably not work.

Only the atoms or molecules in the outer layer of the reactant material can touch the other reactants and react. **Figure 17A** shows that when particles are large, most of the iron atoms are stuck inside and can't react. In **Figure 17B,** more of the reactant atoms are exposed to the oxygen and can react.

Slowing Down Reactions

Sometimes reactions occur too quickly. For example, food and medications can undergo chemical reactions that cause them to spoil or lose their effectiveness too rapidly. Luckily, these reactions can be slowed down.

A substance that slows down a chemical reaction is called an **inhibitor.** An inhibitor makes the formation of a certain amount of product take longer. Some inhibitors completely stop reactions. Many cereals and cereal boxes contain the compound butylated hydroxytoluene, or BHT. The BHT slows the spoiling of the cereal and increases its shelf life.

Figure 18 BHT, an inhibitor, is found in many cereals and cereal boxes.

Speeding Up Reactions

Is it possible to speed up a chemical reaction? Yes, you can add a catalyst (KAT uh lihst). A **catalyst** is a substance that speeds up a chemical reaction. Catalysts do not appear in chemical equations, because they are not changed permanently or used up. A reaction using a catalyst will not produce more product than a reaction without a catalyst, but it will produce the same amount of product faster.

Reading Check *What does a catalyst do in a chemical reaction?*

How does a catalyst work? Many catalysts speed up reaction rates by providing a surface for the reaction to take place. Sometimes the reacting molecules are held in a particular position that favors reaction. Other catalysts reduce the activation energy needed to start the reaction. When the activation energy is reduced, the reaction rate increases.

Catalytic Converters Catalysts are used in the exhaust systems of cars and trucks to aid fuel combustion. The exhaust passes through the catalyst, often in the form of beads coated with metals such as platinum or rhodium. Catalysts speed the reactions that change incompletely burned substances that are harmful, such as carbon monoxide, into less harmful substances, such as carbon dioxide. Similarly, hydrocarbons are changed into carbon dioxide and water. The result of these reactions is cleaner air. These reactions are shown in **Figure 19.**

Breathe Easy The Clean Air Act of 1970 required the reduction of 90 percent of automobile tailpipe emissions. The reduction of emissions included the amount of hydrocarbons and carbon monoxide released. Automakers did not have the technology to meet this new standard. After much hard work, the result of this legislation was the introduction of the catalytic converter in 1975.

Figure 19 Catalytic converters help to complete combustion of fuel. Hot exhaust gases pass over the surfaces of metal-coated beads. On the surface of the beads, carbon monoxide and hydrocarbons are converted to CO_2 and H_2O.

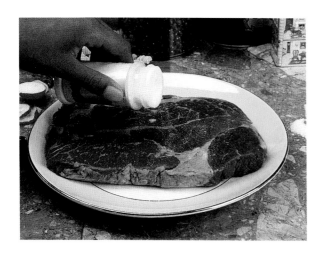

Figure 20 The enzymes in meat tenderizer break down protein in meat, making it more tender.

Enzymes Are Specialists

Some of the most effective catalysts are at work in thousands of reactions that take place in your body. These catalysts, called **enzymes,** are large protein molecules that speed up reactions needed for your cells to work properly. They help your body convert food to fuel, build bone and muscle tissue, convert extra energy to fat, and even produce other enzymes.

These are complex reactions. Without enzymes, they would occur at rates that are too slow to be useful or they would not occur at all. Enzymes make it possible for your body to function. Like other catalysts, enzymes function by positioning the reacting molecules so that their structures fit together properly. Enzymes are a kind of chemical specialist—enzymes exist to carry out each type of reaction in your body.

Other Uses

Enzymes work outside your body, too. One class of enzymes, called proteases (PROH tee ay ses), specializes in protein reactions. They work within cells to break down large, complex molecules called proteins. The meat tenderizer shown in **Figure 20** contains proteases that break down protein in meat, making it more tender. Contact lens cleaning solutions also contain proteases that break down proteins from your eyes that can collect on your lenses and cloud your view.

---

section 2 review

Summary

Chemical Reactions

- To form new bonds in the product, old bonds must break in the reactants. This takes energy.
- Activation energy is the minimum quantity of energy needed to start a reaction.

Reaction Rate

- The rate of reaction tells you how fast a reaction occurs.
- Temperature, concentration, and surface area affect the rate of reaction.

Inhibitors and Catalysts

- Inhibitors slow down reactions. Catalysts speed up reactions.
- Enzymes are catalysts that speed up or slow down reactions for your cells.

Self Check

1. **Describe** how you can measure reaction rates.
2. **Explain** in the general reaction A + B + energy → C, how the following will affect the reaction rate.
   a. increasing the temperature
   b. decreasing the reactant concentration
3. **Describe** how catalysts work to speed up chemical reactions.
4. **Think Critically** Explain why a jar of spaghetti sauce can be stored for weeks on the shelf in the market but must be placed in the refrigerator after it is opened.

Applying Math

5. **Solve One-Step Equations** A chemical reaction is proceeding at a rate of 2 g of product every 45 s. How long will it take to obtain 50 g of product?

Science Online bookl.msscience.com/self_check_quiz

Physical or Chemical Change?

⊙ Real-World Question

Matter can undergo two kinds of changes—physical and chemical. A physical change affects the physical properties. When a chemical change takes place, a new product is produced. How can a scientist tell if a chemical change took place?

Goals

■ **Determine** if a physical or chemical change took place.

Materials

500-mL Erlenmeyer flask
1,000-mL graduated cylinder
one-hole stopper with 15-cm length of glass
    tube inserted
1,000-mL beaker
45-cm length of rubber (or plastic) tubing
stopwatch or clock with second hand
weighing dish   balance
baking soda   vinegar

Safety Precautions

WARNING: *Vinegar (acetic acid) may cause skin and eye irritation.*

⊙ Procedure

1. Measure 300 mL of water. Pour water into 500-mL Erlenmeyer flask.

2. Weigh 5 g of baking soda. Carefully pour the baking soda into the flask. Swirl the flask until the solution is clear.

3. Insert the rubber stopper with the glass tubing into the flask.

4. Measure 600 mL of water and pour into the 1,000-mL beaker.

5. Attach one end of the rubber tubing to the top of the glass tubing. Place the other end of the rubber tubing in the beaker. Be sure the rubber tubing remains under the water.

6. Remove the stopper from the flask. Carefully add 80 mL of vinegar to the flask. Replace the stopper.

7. Count the number of bubbles coming into the beaker for 20 s. Repeat this two more times.

8. Record your data in your Science Journal.

⊙ Conclude and Apply

1. **Describe** what you observed in the flask after the acid was added to the baking soda solution.

2. **Classify** Was this a physical or chemical change? How do you know?

3. **Analyze Results** Was this process endothermic or exothermic?

4. **Calculate** the average reaction rate based on the number of bubbles per second.

𝒞ommunicating Your Data

Compare your results with those of other students in your class.

Design Your Own

Exothermic or Endothermic?

Goals

- **Design** an experiment to test whether a reaction is exothermic or endothermic.
- **Measure** the temperature change caused by a chemical reaction.

Possible Materials

test tubes (8)
test-tube rack
3% hydrogen peroxide solution
raw liver
raw potato
thermometer
stopwatch
clock with second hand
25-mL graduated cylinder

Safety Precautions

WARNING: *Hydrogen peroxide can irritate skin and eyes and damage clothing.* Be careful when handling glass thermometers. Test tubes containing hydrogen peroxide should be placed and kept in racks. Dispose of materials as directed by your teacher. Wash your hands when you complete this lab.

● *Real-World Question*

Energy is always a part of a chemical reaction. Some reactions need energy to start. Other reactions release energy into the environment. What evidence can you find to show that a reaction between hydrogen peroxide and liver or potato is exothermic or endothermic? Think about the difference between these two types of reactions.

● *Form a Hypothesis*

Make a hypothesis that describes how you can use the reactions between hydrogen peroxide and liver or potato to determine whether a reaction is exothermic or endothermic.

● *Test Your Hypothesis*

Make a Plan

1. As a group, look at the list of materials. Decide which procedure you will use to test your hypothesis, and which measurements you will make.

2. **Decide** how you will detect the heat released to the environment during the reaction. Determine how many measurements you will need to make during a reaction.

3. You will get more accurate data if you repeat each experiment several times. Each repeated experiment is called a trial. Use the average of all the trials as your data for supporting your hypothesis.

4. **Decide** what the variables are and what your control will be.

5. **Copy** the data table in your Science Journal before you begin to carry out your experiment.

Follow Your Plan

1. Make sure your teacher approves your plan before you start.

2. Carry out your plan.

3. **Record** your measurements immediately in your data table.

4. **Calculate** the averages of your trial results and record them in your Science Journal.

Analyze Your Data

1. Can you infer that a chemical reaction took place? What evidence did you observe to support this?

2. **Identify** what the variables were in this experiment.

3. **Identify** the control.

Temperature After Adding Liver/Potato				
Trial	Temperature After Adding Liver (°C)		Temperature After Adding Potato (°C)	
	Starting	After____min	Starting	After____min
1				
2		Do not write in this book.		
3				
4				

Conclude and Apply

1. Do your observations allow you to distinguish between an exothermic reaction and an endothermic reaction? Use your data to explain your answer.

2. Where do you think that the energy involved in this experiment came from? Explain your answer.

Communicating Your Data

Compare the results obtained by your group with those obtained by other groups. Are there differences? **Explain** how these might have occurred.

Synthetic Diamonds

Natural Diamond

Almost the Real Thing

Synthetic Diamond

Diamonds are the most dazzling, most dramatic, most valuable natural objects on Earth. Strangely, these beautiful objects are made of carbon, the same material graphite—the stuff found in pencils—is made of. So why is a diamond hard and clear and graphite soft and black? A diamond's hardness is a result of how strongly its atoms are linked. What makes a diamond transparent is the way its crystals are arranged. The carbon in a diamond is almost completely pure, with trace amounts of boron and nitrogen in it. These elements account for the many shades of color found in diamonds.

A diamond is the hardest naturally occurring substance on Earth. It's so hard, only a diamond can scratch another diamond. Diamonds are impervious to heat and household chemicals. Their crystal structure allows them to be split (or crushed) along particular lines.

Diamonds are made when carbon is squeezed at high pressures and temperatures in Earth's upper mantle, about 150 km beneath the surface. At that depth, the temperature is about 1,400°C, and the pressure is about 55,000 atmospheres greater than the pressure at sea level.

As early as the 1850s, scientists tried to convert graphite into diamonds. It wasn't until 1954 that researchers produced the first synthetic diamonds by compressing carbon under extremely high pressure and heat. Scientists converted graphite powder into tiny diamond crystals using pressure of more than 68,000 atm, and a temperature of about 1,700°C for about 16 hours.

Synthetic diamonds are human-made, but they're not fake. They have all the properties of natural diamonds, from hardness to excellent heat conductivity. Experts claim to be able to detect synthetics because they contain tiny amounts of metal (used in their manufacturing process) and have a different luminescence than natural diamonds. In fact, most synthetics are made for industrial use. One major reason is that making small synthetic diamonds is cheaper than finding small natural ones. The other reason is that synthetics can be made to a required size and shape. Still, if new techniques bring down the cost of producing large, gem-quality synthetic diamonds, they may one day compete with natural diamonds as jewelry.

Research Investigate the history of diamonds—natural and synthetic. Explain the differences between them and their uses. Share your findings with the class.

Science online

For more information, visit bookl.msscience.com/time

Reviewing Main Ideas

Section 1 **Formulas and Chemical Equations**

1. Chemical reactions often cause observable changes, such as a change in color or odor, a release or absorption of heat or light, or a release of gas.

2. A chemical equation is a shorthand method of writing what happens in a chemical reaction. Chemical equations use symbols to represent the reactants and products of a reaction, and sometimes show whether energy is produced or absorbed.

3. The law of conservation of mass requires that the same number of atoms of each element be in the products as in the reactants of a chemical equation. This is true in every balanced chemical equation.

Section 2 **Rates of Chemical Reactions**

1. The rate of reaction is a measure of how quickly a reaction occurs.

2. All reactions have an activation energy—a certain minimum amount of energy required to start the reaction.

3. The rate of a chemical reaction can be influenced by the temperature, the concentration of the reactants, and the exposed surface area of the reactant particles.

4. Catalysts can speed up a reaction without being used up. Inhibitors slow down the rate of reaction.

5. Enzymes are protein molecules that act as catalysts in your body's cells.

Visualizing Main Ideas

Copy and complete the following concept map on chemical reactions.

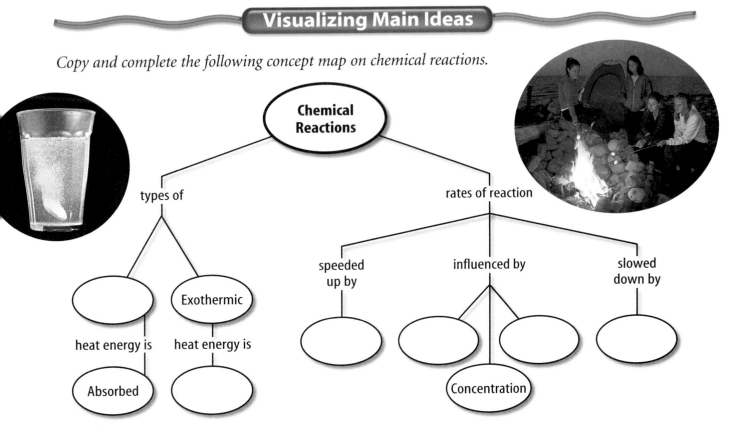

Using Vocabulary

activation energy p.47
catalyst p.51
chemical equation p.192
chemical reaction p.190
concentration p.49
endothermic reaction p.43

enzyme p.52
exothermic reaction p.43
inhibitor p.50
product p.38
rate of reaction p.48
reactant p.38

Explain the differences between the vocabulary terms in each of the following sets.

1. exothermic reaction—endothermic reaction
2. activation energy—rate of reaction
3. reactant—product
4. catalyst—inhibitor
5. concentration—rate of reaction
6. chemical equation—reactant
7. inhibitor—product
8. catalyst—chemical equation
9. rate of reaction—enzyme

Checking Concepts

Choose the word or phrase that best answers the question.

10. Which statement about the law of conservation of mass is NOT true?
    A) The mass of reactants must equal the mass of products.
    B) All the atoms on the reactant side of an equation are also on the product side.
    C) The reaction creates new types of atoms.
    D) Atoms are not lost, but are rearranged.

11. To slow down a chemical reaction, what should you add?
    A) catalyst   C) inhibitor
    B) reactant   D) enzyme

12. Which of these is a chemical change?
    A) Paper is shredded.
    B) Liquid wax turns solid.
    C) A raw egg is broken.
    D) Soap scum forms.

13. Which of these reactions releases heat energy?
    A) unbalanced   C) exothermic
    B) balanced   D) endothermic

14. A balanced chemical equation must have the same number of which of these on both sides of the equation?
    A) atoms   C) molecules
    B) reactants   D) compounds

15. What does NOT affect reaction rate?
    A) balancing   C) surface area
    B) temperature   D) concentration

16. Which is NOT a balanced equation?
    A) $CuCl_2 + H_2S \rightarrow CuS + 2HCl$
    B) $AgNO_3 + NaI \rightarrow AgI + NaNO_3$
    C) $2C_2H_6 + 7O_2 \rightarrow 4CO_2 + 6H_2O$
    D) $MgO + Fe \rightarrow Fe_2O_3 + Mg$

17. Which is NOT evidence that a chemical reaction has occurred?
    A) Milk tastes sour.
    B) Steam condenses on a cold window.
    C) A strong odor comes from a broken egg.
    D) A slice of raw potato darkens.

18. Which of the following would decrease the rate of a chemical reaction?
    A) increase the temperature
    B) reduce the concentration of a reactant
    C) increase the concentration of a reactant
    D) add a catalyst

19. Which of these describes a catalyst?
    A) It is a reactant.
    B) It speeds up a reaction.
    C) It appears in the chemical equation.
    D) It can be used in place of an inhibitor.

Science Online bookl.msscience.com/vocabulary_puzzlemaker

Thinking Critically

20. Cause and Effect Pickled cucumbers remain edible much longer than fresh cucumbers do. Explain.

21. Analyze A beaker of water in sunlight becomes warm. Has a chemical reaction occurred? Explain.

22. Distinguish if $2Ag + S$ is the same as Ag_2S. Explain.

23. Infer Apple slices can be kept from browning by brushing them with lemon juice. Infer what role lemon juice plays in this case.

24. Draw a Conclusion Chili can be made using ground meat or chunks of meat. Which would you choose, if you were in a hurry? Explain.

Use the graph below to answer question 25.

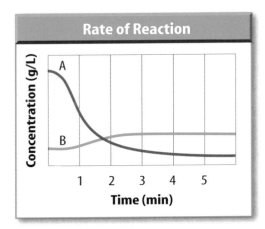

Rate of Reaction

25. Interpret Scientific Illustrations The two curves on the graph represent the concentrations of compounds A (blue) and B (red) during a chemical reaction.
  a. Which compound is a reactant?
  b. Which compound is a product?
  c. During which time period is the concentration of the reactant changing most rapidly?

26. Form a Hypothesis You are cleaning out a cabinet beneath the kitchen sink and find an unused steel wool scrub pad that has rusted completely. Will the remains of this pad weigh more or less than when it was new? Explain.

Performance Activities

27. Poster Make a list of the preservatives in the food you eat in one day. Present your findings to your class in a poster.

Applying Math

Use the graph below to answer question 28.

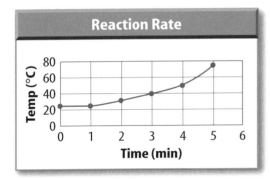

Reaction Rate

28. Reaction Rates In the reaction graph above, how long does it take the reaction to reach $50°C$?

29. Chemical Equation In the following chemical equation, $3Na + AlCl_3 \rightarrow 3NaCl + Al$, how many aluminum molecules will be produced if you have 30 molecules of sodium?

30. Catalysis A zinc catalyst is used to reduce the reaction time by 30%. If the normal time for the reaction to finish is 3 h, how long will it take with the catalyst?

31. Molecules Silver has 6.023×10^{23} molecules per 107.9 g. How many molecules are there if you have
  a. 53.95 g?
  b. 323.7 g?
  c. 10.79 g?

Part 1 Multiple Choice

Record your answers on the answer sheet provided by your teacher or on a sheet of paper.

Use the photo below to answer questions 1 and 2.

1. The photograph shows the reaction of copper (Cu) with silver nitrate ($AgNO_3$) to produce copper nitrate ($Cu(NO_3)_2$) and silver (Ag). The chemical equation that describes this reaction is the following:

$$2AgNO_3 + Cu \rightarrow Cu(NO_3)_2 + 2Ag$$

What term describes what is happening in the reaction?
  A. catalyst
  B. chemical change
  C. inhibitor
  D. physical change

2. Which of the following terms describes the copper on the left side of the equation?
  A. reactant   **C.** enzyme
  B. catalyst   **D.** product

3. Which of the following terms best describes a chemical reaction that absorbs heat energy?
  A. catalytic   **C.** endothermic
  B. exothermic   **D.** acidic

4. What should be balanced in a chemical equation?
  A. electrons   **C.** molecules
  B. atoms   **D.** molecules and atoms

Test-Taking Tip

Read All Questions Never skip a question. If you are unsure of an answer, mark your best guess on another sheet of paper and mark the question in your test booklet to remind you to come back to it at the end of the test.

Use the photo below to answer questions 5 and 6.

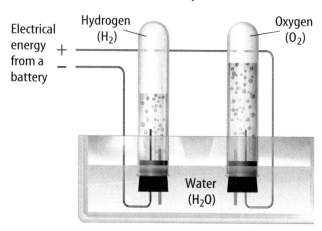

5. The photograph above shows a demonstration of electrolysis, in which water is broken down into hydrogen and oxygen. Which of the following is the best way to write the chemical equation for this process?
  A. $H_2O + energy \rightarrow H_2 + O_2$
  B. $H_2O + energy \rightarrow 2H_2 + O_2$
  C. $2H_2O + energy \rightarrow 2H_2 + O_2$
  D. $2H_2O + energy \rightarrow 2H_2 + 2O_2$

6. For each atom of hydrogen that is present before the reaction begins, how many atoms of hydrogen are present after the reaction?
  A. 1   **C.** 4
  B. 2   **D.** 8

7. What is the purpose of an inhibitor in a chemical reaction?
  A. decrease the shelf life of food
  B. increase the surface area
  C. decrease the speed of a chemical reaction
  D. increase the speed of a chemical reaction

Part 2 | Short Response/Grid In

Record your answers on the answer sheet provided by your teacher or on a sheet of paper.

8. Is a change in the volume of a substance a physical or a chemical change? Explain.

Use the equation below to answer question 9.

$$CaCl_2 + 2AgNO_3 \rightarrow 2 \boxed{} + Ca(NO_3)_2$$

9. When solutions of calcium chloride ($CaCl_2$) and silver nitrate ($AgNO_3$) are mixed, calcium nitrate ($Ca(NO_3)_2$) and a white precipitate, or residue, form. Determine the chemical formula of the precipitate.

Use the illustration below to answer questions 10 and 11.

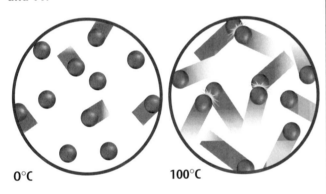

0°C   100°C

10. The figure above demonstrates the movement of water molecules at temperatures of 0°C and 100°C. What would happen to the movement of the molecules if the temperature dropped far below 0°C?

11. Describe how the difference in the movement of the molecules at two different temperatures affects the rate of most chemical reactions.

12. Is activation energy needed for reactions that release energy? Explain why or why not.

Part 3 | Open Ended

Record your answers on a sheet of paper.

Use the illustration below to answer questions 13 and 14.

13. The photograph above shows a forest fire that began when lightning struck a tree. Describe the chemical reaction that occurs when trees burn. Is the reaction endothermic or exothermic? What does this mean? Why does this cause a forest fire to spread?

14. The burning of logs in a forest fire is a chemical reaction. What prevents this chemical reaction from occurring when there is no lightning to start a fire?

15. Explain how the surface area of a material can affect the rate at which the material reacts with other substances. Give an example to support your answer.

16. One of the chemical reactions that occurs in the formation of glass is the combining of calcium carbonate ($CaCO_3$) and silica (SiO_2) to form calcium silicate ($CaSiO_3$) and carbon dioxide (CO_2):

$$CaCO_3 + SiO_2 \rightarrow CaSiO_3 + CO_2$$

Describe this reaction using the names of the chemicals. Discuss which bonds are broken and how atoms are rearranged to form new bonds.

Substances, Mixtures, and Solubility

Big-Band Mixtures

It's a parade and the band plays. Just as the mixing of notes produces music, the mixing of substances produces many of the things around you. From the brass in tubas to the lemonade you drink, you live in a world of mixtures. In this chapter, you'll learn why some substances form mixtures and others do not.

Science Journal Find and name four items around you that are mixtures.

Start-Up Activities

Particle Size and Dissolving Rates

Why do drink mixes come in powder form? What would happen if you dropped a big chunk of drink mix into the water? Would it dissolve quickly? Powdered drink mix dissolves faster in water than chunks do because it is divided into smaller particles, exposing more of the mix to the water. See for yourself how particle size affects the rate at which a substance dissolves.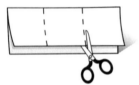

1. Pour 400 mL of water into each of two 600-mL beakers.

2. Carefully grind a bouillon cube into powder using a mortar and pestle.

3. Place the bouillon powder into one beaker and drop a whole bouillon cube into the second beaker.

4. Stir the water in each beaker for 10 s and observe.

5. **Think Critically** Write a paragraph in your Science Journal comparing the color of the two liquids and the amount of undissolved bouillon at the bottom of each beaker. How does the particle size affect the rate at which a substance dissolves?

Solutions Make the following Foldable to help classify solutions based on their common features.

STEP 1 Fold a vertical sheet of paper from side to side. Make the front edge about 1.25 cm shorter than the back edge.

STEP 2 Turn lengthwise and fold into thirds.

STEP 3 Unfold and cut only the top layer along both folds to make three tabs.

STEP 4 Label each tab as shown.

Find Main Ideas As you read the chapter, classify solutions based on their states and list them under the appropriate tabs. On your Foldable, circle the solutions that are acids and underline the solutions that are bases.

Preview this chapter's content and activities at
bookl.msscience.com

section 1

What is a solution?

as you read

What You'll Learn

- **Distinguish** between substances and mixtures.
- **Describe** two different types of mixtures.
- **Explain** how solutions form.
- **Describe** different types of solutions.

Why It's Important

The air you breathe, the water you drink, and even parts of your body are all solutions.

Review Vocabulary

proton: positively charged particle located in the nucleus of an atom

New Vocabulary

- substance
- heterogeneous mixture
- homogeneous mixture
- solution   • solvent
- solute   • precipitate

Substances

Water, salt water, and pulpy orange juice have some obvious differences. These differences can be explained by chemistry. Think about pure water. No matter what you do to it physically—freeze it, boil it, stir it, or strain it—it still is water. On the other hand, if you boil salt water, the water turns to gas and leaves the salt behind. If you strain pulpy orange juice, it loses its pulp. How does chemistry explain these differences? The answer has to do with the chemical compositions of the materials.

Atoms and Elements Recall that atoms are the basic building blocks of matter. Each atom has unique chemical and physical properties which are determined by the number of protons it has. For example, all atoms that have eight protons are oxygen atoms. A **substance** is matter that has the same fixed composition and properties. It can't be broken down into simpler parts by ordinary physical processes, such as boiling, grinding, or filtering. Only a chemical process can change a substance into one or more new substances. **Table 1** lists some examples of physical and chemical processes. An element is an example of a pure substance; it cannot be broken down into simpler substances. The number of protons in an element, like oxygen, are fixed—it cannot change unless the element changes.

Compounds Water is another example of a substance. It is always water even when you boil it or freeze it. Water, however, is not an element. It is an example of a compound which is made of two or more elements that are chemically combined. Compounds also have fixed compositions. The ratio of the atoms in a compound is always the same. For example, when two hydrogen atoms combine with one oxygen atom, water is formed. All water—whether it's in the form of ice, liquid, or steam—has the same ratio of hydrogen atoms to oxygen atoms.

Table 1 Examples of Physical and Chemical Processes	
Physical Processes	**Chemical Processes**
Boiling	Burning
Changing pressure	Reacting with other chemicals
Cooling	Reacting with light
Sorting	

64 ◆ **L CHAPTER 3** Substances, Mixtures, and Solubility

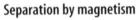
Separation by magnetism

Separation by straining

Figure 1 Mixtures can be separated by physical processes.
Explain *why the iron-sand mixture and the pulpy lemonade are not pure substances.*

Mixtures

Imagine drinking a glass of salt water. You would know right away that you weren't drinking pure water. Like salt water, many things are not pure substances. Salt water is a mixture of salt and water. Mixtures are combinations of substances that are not bonded together and can be separated by physical processes. For example, you can boil salt water to separate the salt from the water. If you had a mixture of iron filings and sand, you could separate the iron filings from the sand with a magnet. **Figure 1** shows some mixtures being separated.

Unlike compounds, mixtures do not always contain the same proportions of the substances that they are composed of. Lemonade is a mixture that can be strong tasting or weak tasting, depending on the amounts of water and lemon juice that are added. It also can be sweet or sour, depending on how much sugar is added. But whether it is strong, weak, sweet, or sour, it is still lemonade.

Heterogeneous Mixtures It is easy to tell that some things are mixtures just by looking at them. A watermelon is a mixture of fruit and seeds. The seeds are not evenly spaced through the whole melon—one bite you take might not have any seeds in it and another bite might have several seeds. A type of mixture where the substances are not mixed evenly is called a **heterogeneous** (he tuh ruh JEE nee us) **mixture.** The different areas of a heterogeneous mixture have different compositions. The substances in a heterogeneous mixture are usually easy to tell apart, like the seeds from the fruit of a watermelon. Other examples of heterogeneous mixtures include a bowl of cold cereal with milk and the mixture of pens, pencils, and books in your backpack.

Science Online

Topic: Desalination
Visit bookl.msscience.com for Web links to information about how salt is removed from salt water to provide drinking water.

Activity Compare and contrast the two most common methods used for desalination.

Sugar

Water

Homogeneous Mixtures Your shampoo contains many ingredients, but you can't see them when you look at the shampoo. It is the same color and texture throughout. Shampoo is an example of a homogeneous (hoh muh JEE nee us) mixture. A **homogeneous mixture** contains two or more substances that are evenly mixed on a molecular level but still are not bonded together. Another name for a homogeneous mixture is a **solution.** The sugar and water in the frozen pops shown in **Figure 2,** are a solution—the sugar is evenly distributed in the water, and you can't see the sugar.

Figure 2 Molecules of sugar and water are evenly mixed in frozen pops.

Reading Check *What is another name for a homogeneous mixture?*

How Solutions Form

How do you make sugar water for a hummingbird feeder? You might add sugar to water and heat the mixture until the sugar disappears. The sugar molecules would spread out until they were evenly spaced throughout the water, forming a solution. This is called dissolving. The substance that dissolves—or seems to disappear—is called the **solute.** The substance that dissolves the solute is called the **solvent.** In the hummingbird feeder solution, the solute is the sugar and the solvent is water. The substance that is present in the greatest quantity is the solvent.

Figure 3 Minerals and soap react to form soap scum, which comes out of the water solution and coats the tiles of a shower.

Forming Solids from Solutions Under certain conditions, a solute can come back out of its solution and form a solid. This process is called crystallization. Sometimes this occurs when the solution is cooled or when some of the solvent evaporates. Crystallization is the result of a physical change. When some solutions are mixed, a chemical reaction occurs, forming a solid. This solid is called a **precipitate** (prih SIH puh tayt). A precipitate is the result of a chemical change. Precipitates probably have formed in your sink or shower because of chemical reactions. Minerals that are dissolved in tap water react chemically with soap. The product of this reaction leaves the water as a precipitate called soap scum, shown in **Figure 3.**

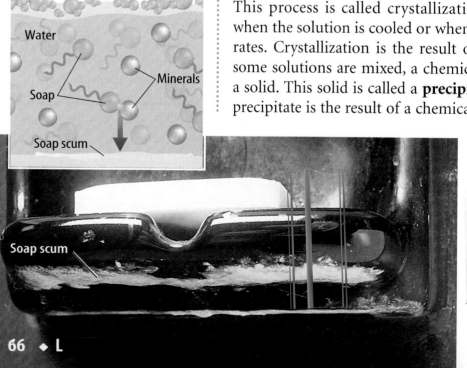

Water

Soap

Minerals

Soap scum

Soap scum

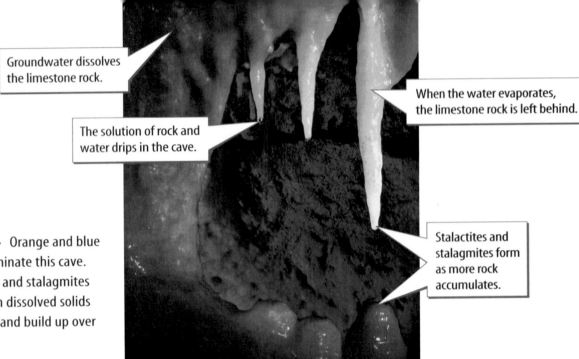

Groundwater dissolves the limestone rock.

When the water evaporates, the limestone rock is left behind.

The solution of rock and water drips in the cave.

Stalactites and stalagmites form as more rock accumulates.

Figure 4 Orange and blue lights illuminate this cave. Stalactites and stalagmites form when dissolved solids crystallize and build up over time.

INTEGRATE
Environment

Stalactites and stalagmites in caves are formed from solutions, as shown in **Figure 4.** First, minerals dissolve in water as it flows through rocks at the top of the cave. This solution of water and dissolved minerals drips from the ceiling of the cave. When drops of the solution evaporate from the roof of the cave, the minerals are left behind. They create the hanging rock formations called stalactites. When drops of the solution fall onto the floor of the cave and evaporate, they form stalagmites. Very often, a stalactite develops downward while a stalagmite develops upward until the two meet. One continuous column of minerals is formed. This process will be discussed later.

Types of Solutions

So far, you've learned about types of solutions in which a solid solute dissolves in a liquid solvent. But solutions can be made up of different combinations of solids, liquids, and gases, as shown in **Table 2.**

Table 2 Examples of Common Solutions

	Solvent/State	Solute/State	State of Solution
Earth's atmosphere	nitrogen/gas	oxygen/gas carbon dioxide/gas argon/gas	gas
Ocean water	water/liquid	salt/solid oxygen/gas carbon dioxide/gas	liquid
Carbonated beverage	water/liquid	carbon dioxide/gas	liquid
Brass	copper/solid	zinc/solid	solid

Figure 5 Acetic acid (a liquid), carbon dioxide (a gas), and drink-mix crystals (a solid) can be dissolved in water (a liquid). **Determine** *whether one liquid solution could contain all three different kinds of solute.*

Liquid Solutions

You're probably most familiar with liquid solutions like the ones shown in **Figure 5,** in which the solvent is a liquid. The solute can be another liquid, a solid, or even a gas. You've already learned about liquid-solid solutions such as sugar water and salt water. When discussing solutions, the state of the solvent usually determines the state of the solution.

Liquid-Gas Solutions Carbonated beverages are liquid-gas solutions—carbon dioxide is the gaseous solute, and water is the liquid solvent. The carbon dioxide gas gives the beverage its fizz and some of its tartness. The beverage also might contain other solutes, such as the compounds that give it its flavor and color.

✔ Reading Check *What are the solutes in a carbonated beverage?*

Liquid-Liquid Solutions In a liquid-liquid solution, both the solvent and the solute are liquids. Vinegar, which you might use to make salad dressing, is a liquid-liquid solution made of 95 percent water (the solvent) and 5 percent acetic acid (the solute).

Gaseous Solutions

In gaseous solutions, a smaller amount of one gas is dissolved in a larger amount of another gas. This is called a gas-gas solution because both the solvent and solute are gases. The air you breathe is a gaseous solution. Nitrogen makes up about 78 percent of dry air and is the solvent. The other gases are the solutes.

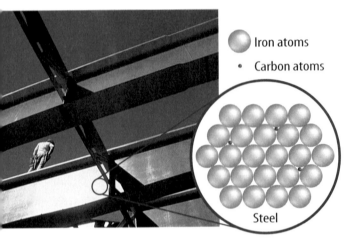

Figure 6 Metal alloys can contain either metal or nonmetal solutes dissolved in a metal solvent.

Copper atoms

Zinc atoms

Iron atoms

• Carbon atoms

Brass

Steel

Steel is a solid solution of the metal iron and the nonmetal carbon.

Brass is a solid solution made of copper and zinc.

Solid Solutions In solid solutions, the solvent is a solid. The solute can be a solid, liquid, or gas. The most common solid solutions are solid-solid solutions—ones in which the solvent and the solute are solids. A solid-solid solution made from two or more metals is called an alloy. It's also possible to include elements that are not metals in alloys. For example, steel is an alloy that has carbon dissolved in iron. The carbon makes steel much stronger and yet more flexible than iron. Two alloys are shown in **Figure 6.**

---

section 1 review

Summary

Substances

● Elements are substances that cannot be broken down into simpler substances.

● A compound is made up of two or more elements bonded together.

Mixtures and Solutions

● Mixtures are either heterogeneous or homogeneous.

● Solutions have two parts—solute and solvent.

● Crystallization and precipitation are two ways that solids are formed from solutions.

Types of Solutions

● The solutes and solvents can be solids, liquids, or gases.

Self Check

1. **Compare and contrast** substances and mixtures. Give two examples of each.

2. **Describe** how heterogeneous and homogeneous mixtures differ.

3. **Explain** how a solution forms.

4. **Identify** the common name for a solid-solid solution of metals.

5. **Think Critically** The tops of carbonated-beverage cans usually are made with a different aluminum alloy than the pull tabs are made with. Explain.

Applying Skills

6. **Compare and contrast** the following solutions: a helium-neon laser, bronze (a copper-tin alloy), cloudy ice cubes, and ginger ale.

Solubility

as you read

What You'll Learn

- **Explain** why water is a good general solvent.
- **Describe** how the structure of a compound affects which solvents it dissolves in.
- **Identify** factors that affect how much of a substance will dissolve in a solvent.
- **Describe** how temperature affects reaction rate.
- **Explain** how solute particles affect physical properties of water.

Why It's Important

How you wash your hands, clothes, and dishes depends on which substances can dissolve in other substances.

Review Vocabulary

polar bond: a bond resulting from the unequal sharing of electrons

New Vocabulary

- aqueous
- saturated
- solubility
- concentration

Water—The Universal Solvent

In many solutions, including fruit juice and vinegar, water is the solvent. A solution in which water is the solvent is called an **aqueous** (A kwee us) solution. Because water can dissolve so many different solutes, chemists often call it the universal solvent. To understand why water is such a great solvent, you must first know a few things about atoms and bonding.

Molecular Compounds When certain atoms form compounds, they share electrons. Sharing electrons is called covalent bonding. Compounds that contain covalent bonds are called molecular compounds, or molecules.

If a molecule has an even distribution of electrons, like the one in **Figure 7,** it is called nonpolar. The atoms in some molecules do not have an even distribution of electrons. For example, in a water molecule, two hydrogen atoms share electrons with a single oxygen atom. However, as **Figure 7** shows, the electrons spend more time around the oxygen atom than they spend around the hydrogen atoms. As a result, the oxygen portion of the water molecule has a partial negative charge and the hydrogen portions have a partial positive charge. The overall charge of the water molecule is neutral. Such a molecule is said to be polar, and the bonds between its atoms are called polar covalent bonds.

Figure 7 Some atoms share electrons to form covalent bonds.

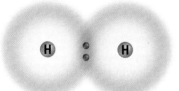

Two atoms of hydrogen share their electrons equally. Such a molecule is nonpolar.

The electrons spend more time around the oxygen atom than the hydrogen atoms. Such a molecule is polar.

(Partial negative charge)

(Partial positive charge)

Ionic Bonds Some atoms do not share electrons when they join with other atoms to form compounds. Instead, these atoms lose or gain electrons. When they do, the number of protons and electrons within an atom are no longer equal, and the atom becomes positively or negatively charged. Atoms with a charge are called ions. Bonds between ions that are formed by the transfer of electrons are called ionic bonds, and the compound that is formed is called an ionic compound. Table salt is an ionic compound that is made of sodium ions and chloride ions. Each sodium atom loses one electron to a chlorine atom and becomes a positively charged sodium ion. Each chlorine atom gains one electron from a sodium atom, becoming a negatively charged chloride ion.

 **Reading Check** *How does an ionic compound differ from a molecular compound?*

How Water Dissolves Ionic Compounds Now think about the properties of water and the properties of ionic compounds as you visualize how an ionic compound dissolves in water. Because water molecules are polar, they attract positive and negative ions. The more positive part of a water molecule—where the hydrogen atoms are—is attracted to negatively charged ions. The more negative part of a water molecule—where the oxygen atom is—attracts positive ions. When an ionic compound is mixed with water, the different ions of the compound are pulled apart by the water molecules. **Figure 8** shows how sodium chloride dissolves in water.

Solutions Seawater is a solution that contains nearly every element found on Earth. Most elements are present in tiny quantities. Sodium and chloride ions are the most common ions in seawater. Several gases, including oxygen, nitrogen, and carbon dioxide, also are dissolved in seawater.

Figure 8 Water dissolves table salt because its partial charges are attracted to the charged ions in the salt.

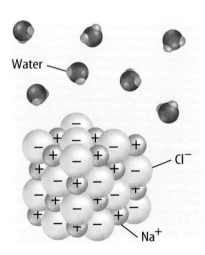

The partially negative oxygen in the water molecule is attracted to a positive sodium ion.

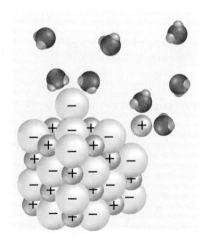

The partially positive hydrogen atoms in another water molecule are attracted to a negative chloride ion.

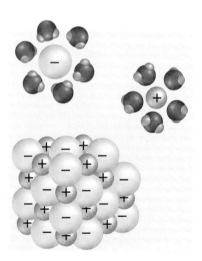

The sodium and chloride ions are pulled apart from each other, and more water molecules are attracted to them.

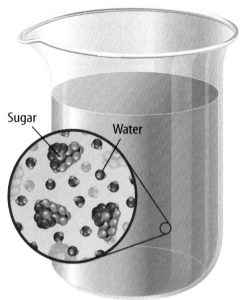

Figure 9 Sugar molecules that are dissolved in water spread out until they are spaced evenly in the water.

Figure 10 Water and oil do not mix because water molecules are polar and oil molecules are nonpolar.

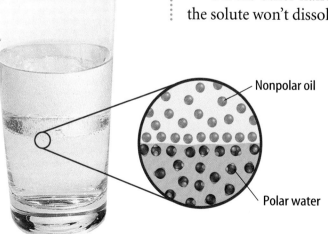

Nonpolar oil

Polar water

How Water Dissolves Molecular Compounds

Can water also dissolve molecular compounds that are not made of ions? Water does dissolve molecular compounds, such as sugar, although it doesn't break each sugar molecule apart. Water simply moves between different molecules of sugar, separating them. Like water, a sugar molecule is polar. Polar water molecules are attracted to the positive and negative portions of the polar sugar molecules. When the sugar molecules are separated by the water and spread throughout it, as **Figure 9** shows, they have dissolved.

What will dissolve?

When you stir a spoonful of sugar into iced tea, all of the sugar dissolves but none of the metal in the spoon does. Why does sugar dissolve in water, but metal does not? A substance that dissolves in another is said to be soluble in that substance. You would say that the sugar is soluble in water but the metal of the spoon is insoluble in water, because it does not dissolve readily.

Like Dissolves Like When trying to predict which solvents can dissolve which solutes, chemists use the rule of "like dissolves like." This means that polar solvents dissolve polar solutes and nonpolar solvents dissolve nonpolar solutes. In the case of sugar and water, both are made up of polar molecules, so sugar is soluble in water. In the case of salt and water, the sodium and chloride ion pair is like the water molecule because it has a positive charge at one end and a negative charge at the other end.

✔ **Reading Check** *What does "like dissolves like" mean?*

On the other hand, if a solvent and a solute are not similar, the solute won't dissolve. For example, oil and water do not mix. Oil molecules are nonpolar, so polar water molecules are not attracted to them. If you pour vegetable oil into a glass of water, the oil and the water separate into layers instead of forming a solution, as shown in **Figure 10.** You've probably noticed the same thing about the oil-and-water mixtures that make up some salad dressings. The oil stays on the top. Oils generally dissolve better in solvents that have nonpolar molecules.

How much will dissolve?

Even though sugar is soluble in water, if you tried to dissolve 1 kg of sugar into one small glass of water, not all of the sugar would dissolve. **Solubility** (sahl yuh BIH luh tee) is a measurement that describes how much solute dissolves in a given amount of solvent. The solubility of a material has been described as the amount of the material that can dissolve in 100 g of solvent at a given temperature. Some solutes are highly soluble, meaning that a large amount of solute can be dissolved in 100 g of solvent. For example, 63 g of potassium chromate can be dissolved in 100 g of water at 25°C. On the other hand, some solutes are not very soluble. For example, only 0.00025 g of barium sulfate will dissolve in 100 g of water at 25°C. When a substance has an extremely low solubility, like barium sulfate does in water, it usually is considered insoluble.

 **Reading Check** *What is an example of a substance that is considered to be insoluble in water?*

Figure 11 The solubility of some solutes changes as the temperature of the solvent increases.
Use a Graph *According to the graph, is it likely that warm ocean water contains any more sodium chloride than cold ocean water does?*

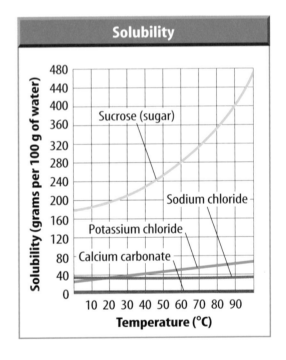

Solubility in Liquid-Solid Solutions Did you notice that the temperature was included in the explanation about the amount of solute that dissolves in a quantity of solvent? The solubility of many solutes changes if you change the temperature of the solvent. For example, if you heat water, not only does the sugar dissolve at a faster rate, but more sugar can dissolve in it. However, some solutes, like sodium chloride and calcium carbonate, do not become more soluble when the temperature of water increases. The graph in **Figure 11** shows how the temperature of the solvent affects the solubility of some solutes.

Solubility in Liquid-Gas Solutions Unlike liquid-solid solutions, an increase in temperature decreases the solubility of a gas in a liquid-gas solution. You might notice this if you have ever opened a warm carbonated beverage and it bubbled up out of control while a chilled one barely fizzed. Carbon dioxide is less soluble in a warm solution. What keeps the carbon dioxide from bubbling out when it is sitting at room temperature on a supermarket shelf? When a bottle is filled, extra carbon dioxide gas is squeezed into the space above the liquid, increasing the pressure in the bottle. This increased pressure increases the solubility of gas and forces most of it into the solution. When you open the cap, the pressure is released and the solubility of the carbon dioxide decreases.

 **Reading Check** *Why does a bottle of carbonated beverage go "flat" after it has been opened for a few days?*

Observing Chemical Processes

Procedure

1. Pour **two small glasses of milk.**
2. Place one glass of milk in the **refrigerator.** Leave the second glass on the counter.
3. Allow the milk to sit overnight. **WARNING:** *Do not drink the milk that sat out overnight.*
4. On the following day, smell both glasses of milk. Record your observations.

Analysis

1. Compare and contrast the smell of the refrigerated milk to the non-refrigerated milk.
2. Explain why refrigeration is needed.

Try at Home

Figure 12 The Dead Sea has an extremely high concentration of dissolved minerals. When the water evaporates, the minerals are left behind and form pillars.

Saturated Solutions If you add calcium carbonate to 100 g of water at 25°C, only 0.0014 g of it will dissolve. Additional calcium carbonate will not dissolve. Such a solution—one that contains all of the solute that it can hold under the given conditions—is called a **saturated** solution. **Figure 12** shows a saturated solution. If a solution is a liquid-solid solution, the extra solute that is added will settle to the bottom of the container. It's possible to make solutions that have less solute than they would need to become saturated. Such solutions are unsaturated. An example of an unsaturated solution is one containing 50 g of sugar in 100 g of water at 25°C. That's much less than the 204 g of sugar the solution would need to be saturated.

A hot solvent usually can hold more solute than a cool solvent can. When a saturated solution cools, some of the solute usually falls out of the solution. But if a saturated solution is cooled slowly, sometimes the excess solute remains dissolved for a period of time. Such a solution is said to be supersaturated, because it contains more than the normal amount of solute.

Rate of Dissolving

Solubility does not tell you how fast a solute will dissolve—it tells you only how much of a solute will dissolve at a given temperature. Some solutes dissolve quickly, but others take a long time to dissolve. A solute dissolves faster when the solution is stirred or shaken or when the temperature of the solution is increased. These methods increase the rate at which the surfaces of the solute come into contact with the solvent. Increasing the area of contact between the solute and the solvent can also increase the rate of dissolving. This can be done by breaking up the solute into smaller pieces, which increases the surface area of the solute that is exposed to the solvent.

INTEGRATE Chemistry

Molecules are always moving and colliding. The collisions must take place for chemical processes to occur. The chemical processes take place at a given rate of reaction. Temperature has a large effect on that rate. The higher the temperature, the more collisions occur and the higher the rate of reaction. The opposite is also true. The lower the temperature, the less collisions occur and the lower the rate of reaction. Refrigerators are an example of slowing the reaction rate—and therefore the chemical process—down to prevent food spoilage.

Concentration

What makes strong lemonade strong and weak lemonade weak? The difference between the two drinks is the amount of water in each one compared to the amount of lemon. The lemon is present in different concentrations in the solution. The **concentration** of a solution tells you how much solute is present compared to the amount of solvent. You can give a simple description of a solution's concentration by calling it either concentrated or dilute. These terms are used when comparing the concentrations of two solutions with the same type of solute and solvent. A concentrated solution has more solute per given amount of solvent than a dilute solution.

Measuring Concentration Can you imagine a doctor ordering a dilute intravenous, or IV, solution for a patient? Because dilute is not an exact measurement, the IV could be made with a variety of amounts of medicine. The doctor would need to specify the exact concentration of the IV solution to make sure that the patient is treated correctly.

Pharmacist Doctors rely on pharmacists to formulate IV solutions. Pharmacists begin with a concentrated form of the drug, which is supplied by pharmaceutical companies. This is the solute of the IV solution. The pharmacist adds the correct amount of solvent to a small amount of the solute to achieve the concentration requested by the doctor. There may be more than one solute per IV solution in varying concentrations.

Applying Science

How can you compare concentrations?

A solute is a substance that can be dissolved in another substance called a solvent. Solutions vary in concentration, or strength, depending on the amount of solute and solvent being used. Fruit drinks are examples of such a solution. Stronger fruit drinks appear darker in color and are the result of more drink mix being dissolved in a given amount of water. What would happen if more water were added to the solution?

Glucose Solutions (g/100 mL)		
Solute Glucose (g)	Solvent Water (mL)	Solution Concentration of Glucose (%)
2	100	2
4	100	4
10	100	10
20	100	20

Identifying the Problem

The table on the right lists different concentration levels of glucose solutions, a type of carbohydrate your body uses as a source of energy. The glucose is measured in grams, and the water is measured in milliliters.

Solving the Problem

A physician writes a prescription for a patient to receive 1,000 mL of a 20 percent solution of glucose. How many grams of glucose must the pharmacist add to 1,000 mL of water to prepare this 20 percent concentration level?

Figure 13 Concentrations can be stated in percentages.
Identify *the percentage of this fruit drink that is water, assuming there are no other dissolved substances.*

One way of giving the exact concentration is to state the percentage of the volume of the solution that is made up of solute. Labels on fruit drinks show their concentration like the one in **Figure 13.** When a fruit drink contains 15 percent fruit juice, the remaining 85 percent of the drink is water and other substances such as sweeteners and flavorings. This drink is more concentrated than another brand that contains 10 percent fruit juice, but it's more dilute than pure juice, which is 100 percent juice. Another way to describe the concentration of a solution is to give the percentage of the total mass that is made up of solute.

Effects of Solute Particles All solute particles affect the physical properties of the solvent, such as its boiling point and freezing point. The effect that a solute has on the freezing or boiling point of a solvent depends on the number of solute particles.

When a solvent such as water begins to freeze, its molecules arrange themselves in a particular pattern. Adding a solute such as sodium chloride to this solvent changes the way the molecules arrange themselves. To overcome this interference of the solute, a lower temperature is needed to freeze the solvent.

When a solvent such as water begins to boil, the solvent molecules are gaining enough energy to move from the liquid state to the gaseous state. When a solute such as sodium chloride is added to the solvent, the solute particles interfere with the evaporation of the solvent particles. More energy is needed for the solvent particles to escape from the liquid, and the boiling point of the solution will be higher.

section 2 review

Summary

The Universal Solvent
- Water is known as the universal solvent.
- A molecule that has an even distribution of electrons is a nonpolar molecule.
- A molecule that has an uneven distribution of electrons is a polar molecule.
- A compound that loses or gains electrons is an ionic compound.

Dissolving a Substance
- Chemists use the rule "like dissolves like."

Concentration
- Concentration is the quantity of solute present compared to the amount of solvent.

Self Check

1. **Identify** the property of water that makes it the universal solvent.
2. **Describe** the two methods to increase the rate at which a substance dissolves.
3. **Infer** why it is important to add sodium chloride to water when making homemade ice cream.
4. **Think Critically** Why can the fluids used to dry-clean clothing remove grease even when water cannot?

Applying Skills

5. **Recognize Cause and Effect** Why is it more important in terms of reaction rate to take groceries straight home from the store when it is 25°C than when it is 2°C?

Science Online   bookl.msscience.com/self_check_quiz

Observing Gas Solubility

On a hot day, a carbonated beverage will cool you off. If you leave the beverage uncovered at room temperature, it quickly loses its fizz. However, if you cap the beverage and place it in the refrigerator, it will still have its fizz hours later. In this lab you will explore why this happens.

▶ Real-World Question

What effect does temperature have on the fizz, or carbon dioxide, in your carbonated beverage?

Goals

■ **Observe** the effect that temperature has on solubility.

■ **Compare** the amount of carbon dioxide released at room temperature and in hot tap water.

Materials

carbonated beverages in plastic bottles, thoroughly chilled (2)

balloons (2)   *ruler
tape   container
fabric tape measure   hot tap water
*string   *Alternative materials

Safety Precautions

WARNING: *DO NOT point the bottles at anyone at any time during the lab.*

▶ Procedure

1. Carefully remove the caps from the thoroughly chilled plastic bottles one at a time. Create as little agitation as possible.

2. Quickly cover the opening of each bottle with an uninflated balloon.

3. Use tape to secure and tightly seal the balloons to the top of the bottles.

4. Gently agitate one bottle from side to side for two minutes. Measure the circumference of the balloon.

WARNING: *Contents under pressure can cause serious accidents. Be sure to wear safety goggles, and DO NOT point the bottles at anyone.*

5. Gently agitate the second bottle in the same manner as in step 4. Then, place the bottle in a container of hot tap water for ten minutes. Measure the circumference of the balloon.

▶ Conclude and Apply

1. **Contrast** the relative amounts of carbon dioxide gas released from the cold and the warm carbonated beverages.

2. **Infer** Why does the warmed carbonated beverage release a different amount of carbon dioxide than the chilled one?

*C*ommunicating
Your Data

Compare the circumferences of your balloons with those of members of your class. **For more help, refer to the** Science Skill Handbook.

Acidic and Basic Solutions

What You'll Learn

- **Compare** acids and bases and their properties.
- **Describe** practical uses of acids and bases.
- **Explain** how pH is used to describe the strength of an acid or base.
- **Describe** how acids and bases react when they are brought together.

Why It's Important

Many common products, such as batteries and bleach, work because of acids or bases.

Review Vocabulary

physical property: any characteristic of a material that can be seen or measured without changing the material

New Vocabulary

- acid
- hydronium ion
- base
- pH
- indicator
- neutralization

Acids

What makes orange juice, vinegar, dill pickles, and grapefruit tangy? Acids cause the sour taste of these and other foods. **Acids** are substances that release positively charged hydrogen ions, H^+, in water. When an acid mixes with water, the acid dissolves, releasing a hydrogen ion. The hydrogen ion then combines with a water molecule to form a hydronium ion, as shown in **Figure 14. Hydronium ions** are positively charged and have the formula H_3O^+.

Properties of Acidic Solutions Sour taste is one of the properties of acidic solutions. The taste allows you to detect the presence of acids in your food. However, even though you can identify acidic solutions by their sour taste, you should never taste anything in the laboratory, and you should never use taste to test for the presence of acids in an unknown substance. Many acids can cause serious burns to body tissues.

Another property of acidic solutions is that they can conduct electricity. The hydronium ions in an acidic solution can carry the electric charges in a current. This is why some batteries contain an acid. Acidic solutions also are corrosive, which means they can break down certain substances. Many acids can corrode fabric, skin, and paper. The solutions of some acids also react strongly with certain metals. The acid-metal reaction forms metallic compounds and hydrogen gas, leaving holes in the metal in the process.

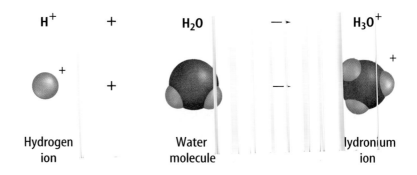

Figure 14 One hydrogen ion can combine with one water molecule to form one positively charged hydronium ion.
Identify *what kinds of substances are sources of hydrogen ions.*

Figure 15 Each of these products contains an acid or is made with the help of an acid.
Describe how your life would be different if acids were not available to make these products.

Uses of Acids You're probably familiar with many acids. Vinegar, which is used in salad dressing, contains acetic acid. Lemons, limes, and oranges have a sour taste because they contain citric acid. Your body needs ascorbic acid, which is vitamin C. Ants that sting inject formic acid into their victims.

Figure 15 shows other products that are made with acids. Sulfuric acid is used in the production of fertilizers, steel, paints, and plastics. Acids often are used in batteries because their solutions conduct electricity. For this reason, it sometimes is referred to as battery acid. Hydrochloric acid, which is known commercially as muriatic acid, is used in a process called pickling. Pickling is a process that removes impurities from the surfaces of metals. Hydrochloric acid also can be used to clean mortar from brick walls. Nitric acid is used in the production of fertilizers, dyes, and plastics.

Acid in the Environment Carbonic acid plays a key role in the formation of caves and of stalactites and stalagmites. Carbonic acid is formed when carbon dioxide in soil is dissolved in water. When this acidic solution comes in contact with calcium carbonate—or limestone rock—it can dissolve it, eventually carving out a cave in the rock. A similar process occurs when acid rain falls on statues and eats away at the stone, as shown in **Figure 16.** When this acidic solution drips from the ceiling of the cave, water evaporates and carbon dioxide becomes less soluble, forcing it out of solution. The solution becomes less acidic and the limestone becomes less soluble, causing it to come out of solution. These solids form stalactites and stalagmites.

Mini LAB

Observing a Nail in a Carbonated Drink

Procedure
1. Observe the initial appearance of an **iron nail.**
2. Pour enough **carbonated soft drink** into a **cup or beaker** to cover the nail.
3. Drop the nail into the soft drink and observe what happens.
4. Leave the nail in the soft drink overnight and observe it again the next day.

Analysis
1. Describe what happened when you first dropped the nail into the soft drink and the appearance of the nail the following day.
2. Based upon the fact that the soft drink was carbonated, explain why you think the drink reacted with the nail as you observed.

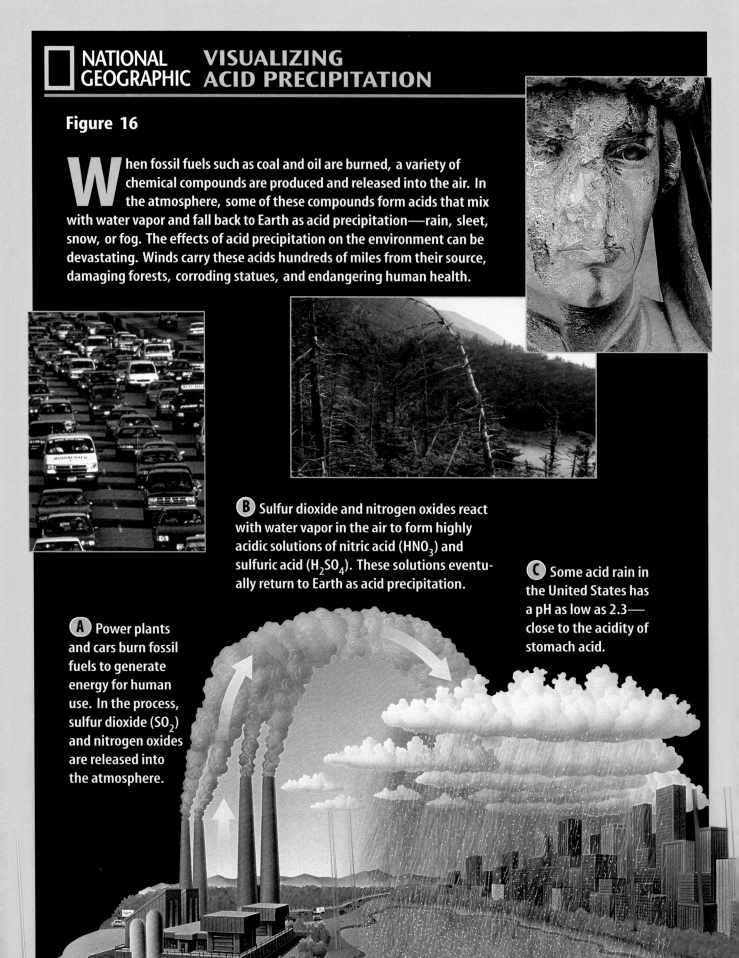

VISUALIZING ACID PRECIPITATION

Figure 16

When fossil fuels such as coal and oil are burned, a variety of chemical compounds are produced and released into the air. In the atmosphere, some of these compounds form acids that mix with water vapor and fall back to Earth as acid precipitation—rain, sleet, snow, or fog. The effects of acid precipitation on the environment can be devastating. Winds carry these acids hundreds of miles from their source, damaging forests, corroding statues, and endangering human health.

B Sulfur dioxide and nitrogen oxides react with water vapor in the air to form highly acidic solutions of nitric acid (HNO_3) and sulfuric acid (H_2SO_4). These solutions eventually return to Earth as acid precipitation.

C Some acid rain in the United States has a pH as low as 2.3—close to the acidity of stomach acid.

A Power plants and cars burn fossil fuels to generate energy for human use. In the process, sulfur dioxide (SO_2) and nitrogen oxides are released into the atmosphere.

Bases

People often use ammonia solutions to clean windows and floors. These solutions have different properties from those of acidic solutions. Ammonia is called a base. **Bases** are substances that can accept hydrogen ions. When bases dissolve in water, some hydrogen atoms from the water molecules are attracted to the base. A hydrogen atom in the water molecule leaves behind the other hydrogen atom and oxygen atom. This pair of atoms is a negatively charged ion called a hydroxide ion. A hydroxide ion has the formula OH^-. Most bases contain a hydroxide ion, which is released when the base dissolves in water. For example, sodium hydroxide is a base with the formula NaOH. When NaOH dissolves in water, a sodium ion and the hydroxide ion separate.

Properties of Basic Solutions Most soaps are bases, so if you think about how soap feels, you can figure out some of the properties of basic solutions. Basic solutions feel slippery. Acids in water solution taste sour, but bases taste bitter—as you know if you have ever accidentally gotten soap in your mouth.

Like acids, bases are corrosive. Bases can cause burns and damage tissue. You should never touch or taste a substance to find out whether it is a base. Basic solutions contain ions and can conduct electricity. Basic solutions are not as reactive with metals as acidic solutions are.

Uses of Bases Many uses for bases are shown in **Figure 17.** Bases give soaps, ammonia, and many other cleaning products some of their useful properties. The hydroxide ions produced by bases can interact strongly with certain substances, such as dirt and grease.

Chalk and oven cleaner are examples of familiar products that contain bases. Your blood is a basic solution. Calcium hydroxide, often called lime, is used to mark the lines on athletic fields. It also can be used to treat lawns and gardens that have acidic soil. Sodium hydroxide, known as lye, is a strong base that can cause burns and other health problems. Lye is used to make soap, clean ovens, and unclog drains.

Science Online

Topic: Calcium Hydroxide
Visit bookl.msscience.com for Web links to information about the uses for calcium hydroxide.

Activity Describe the chemical reaction that converts limestone (calcium carbonate) to calcium hydroxide.

Figure 17 Many products, including soaps, cleaners, and plaster contain bases or are made with the help of bases.

pH Levels Most life-forms can't exist at extremely low pH levels. However, some bacteria thrive in acidic environments. Acidophils are bacteria that exist at low pH levels. These bacteria have been found in the Hot Springs of Yellowstone National Park in areas with pH levels ranging from 1 to 3.

What is pH?

You've probably heard of pH-balanced shampoo or deodorant, and you might have seen someone test the pH of the water in a swimming pool. **pH** is a measure of how acidic or basic a solution is. The pH scale ranges from 0 to 14. Acidic solutions have pH values below 7. A solution with a pH of 0 is very acidic. Hydrochloric acid can have a pH of 0. A solution with a pH of 7 is neutral, meaning it is neither acidic nor basic. Pure water is neutral. Basic solutions have pH values above 7. A solution with a pH of 14 is very basic. Sodium hydroxide can have a pH of 14. **Figure 18** shows where various common substances fall on the pH scale.

The pH of a solution is related directly to its concentrations of hydronium ions (H_3O^+) and hydroxide ions (OH^-). Acidic solutions have more hydronium ions than hydroxide ions. Neutral solutions have equal numbers of the two ions. Basic solutions have more hydroxide ions than hydronium ions.

Reading Check *In a neutral solution, how do the numbers of hydronium ions and hydroxide ions compare?*

pH Scale The pH scale is not a simple linear scale like mass or volume. For example, if one book has a mass of 2 kg and a second book has a mass of 1 kg, the mass of the first book is twice that of the second. However, a change of 1 pH unit represents a tenfold change in the acidity of the solution. For example, if one solution has a pH of 1 and a second solution has a pH of 2, the first solution is not twice as acidic as the second—it is ten times more acidic. To determine the difference in pH strength, use the following calculation: 10^n, where $n =$ the difference between pHs. For example: pH3 − pH1 = 2, $10^2 = 100$ times more acidic.

Figure 18 The pH scale classifies a solution as acidic, basic, or neutral.

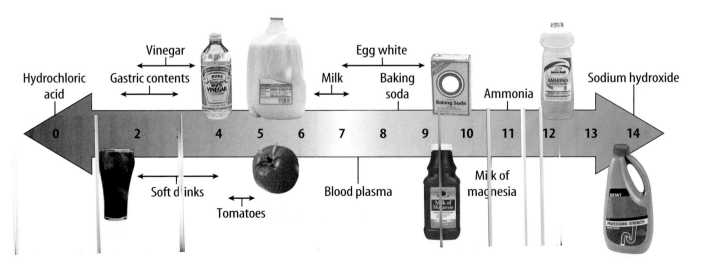

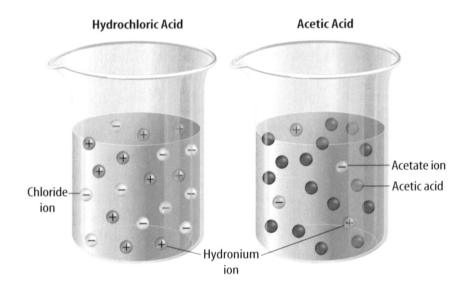

Hydrochloric Acid

Acetic Acid

Chloride ion

Hydronium ion

Acetate ion

Acetic acid

Figure 19 Hydrochloric acid separates into ions more readily than acetic acid does when it dissolves in water. Therefore, hydrochloric acid exists in water as separated ions. Acetic acid exists in water almost entirely as molecules.

Strengths of Acids and Bases You've learned that acids give foods a sour taste but also can cause burns and damage tissue. The difference between food acids and the acids that can burn you is that they have different strengths. The acids in food are fairly weak acids, while the dangerous acids are strong acids. The strength of an acid is related to how easily the acid separates into ions, or how easily a hydrogen ion is released, when the acid dissolves in water. Look at **Figure 19.** In the same concentration, a strong acid—like hydrochloric acid—forms more hydronium ions in solution than a weak acid does—like acetic acid. More hydronium ions means the strong-acid solution has a lower pH than the weak-acid solution. Similarly, the strength of a base is related to how easily the base separates into ions, or how easily a hydroxide ion is released, when the base dissolves in water. The relative strengths of some common acids and bases are shown in **Table 3.**

Reading Check *What determines the strength of an acid or a base?*

An acid containing more hydrogen atoms, such as carbonic acid, H_2CO_3, is not necessarily stronger than an acid containing fewer hydrogen atoms, such as nitric acid, HNO_3. An acid's strength is related to how easily a hydrogen ion separates—not to how many hydrogen atoms it has. For this reason, nitric acid is stronger than carbonic acid.

Table 3 Strengths of Some Acids and Bases		
	Acid	**Base**
Strong	hydrochloric (HCl) sulfuric (H_2SO_4) nitric (HNO_3)	sodium hydroxide (NaOH) potassium hydroxide (KOH)
Weak	acetic (CH_3COOH) carbonic (H_2CO_3) ascorbic ($H_2C_6H_6O_6$)	ammonia (NH_3) aluminum hydroxide ($Al(OH)_3$) iron (III) hydroxide ($Fe(OH)_3$)

Indicators

What is a safe way to find out how acidic or basic a solution is? **Indicators** are compounds that react with acidic and basic solutions and produce certain colors, depending on the solution's pH.

Because they are different colors at different pHs, indicators can help you determine the pH of a solution. Some indicators, such as litmus, are soaked into paper strips. When litmus paper is placed in an acidic solution, it turns red. When placed in a basic solution, litmus paper turns blue. Some indicators can change through a wide range of colors, with each different color appearing at a different pH value.

Neutralization

Perhaps you've heard someone complain about heartburn or an upset stomach after eating spicy food. To feel better, the person might have taken an antacid. Think about the word *antacid* for a minute. How do antacids work?

Heartburn or stomach discomfort is caused by excess hydrochloric acid in the stomach. Hydrochloric acid helps break down the food you eat, but too much of it can irritate your stomach or digestive tract. An antacid product, often made from the base magnesium hydroxide, $Mg(OH)_2$, neutralizes the excess acid. **Neutralization** (new truh luh ZAY shun) is the reaction of an acid with a base. It is called this because the properties of both the acid and base are diminished, or neutralized. In most cases, the reaction produces a water and a salt. **Figure 20** illustrates the relative amounts of hydronium and hydroxide ions between pH 0 and pH 14.

> ✔ **Reading Check** *What are the products of neutralization?*

Science Online

Topic: Indicators

Visit bookl.msscience.com for Web links to information about the types of pH indicators.

Activity Describe how plants can act as indicators in acidic and basic solutions.

Figure 20 The pH of a solution is more acidic when greater amounts of hydronium ions are present.
Define *what makes a pH 7 solution neutral.*

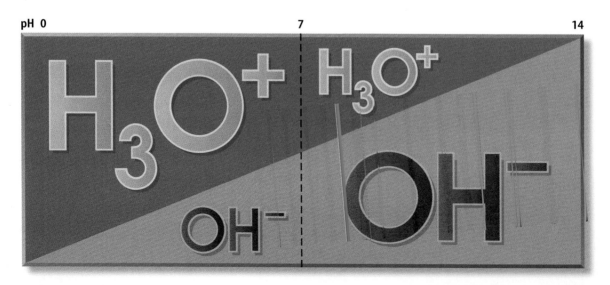

pH 0   7   14

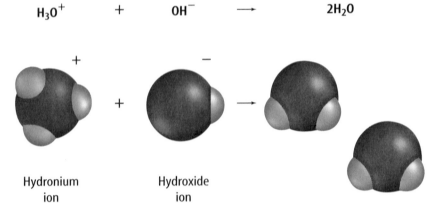

$$H_3O^+ \quad + \quad OH^- \quad \longrightarrow \quad 2H_2O$$

+

+

Hydronium
ion

Hydroxide
ion

Water molecules

Figure 21 When acidic and basic solutions react, hydronium and hydroxide ions react to form water.
Determine *why the pH of the solution changes.*

How does neutralization occur? Recall that every water molecule contains two hydrogen atoms and one oxygen atom. As **Figure 21** shows, when one hydronium ion reacts with one hydroxide ion, the product is two water molecules. This reaction occurs during acid-base neutralization. Equal numbers of hydronium ions from the acidic solution and hydroxide ions from the basic solution react to produce water. Pure water has a pH of 7, which means that it's neutral.

 **Reading Check** *What happens to acids and bases during neutralization?*

section 3 review

Summary

Acids and Bases

- Acids are substances that release positively charged hydrogen ions in water.
- Substances that accept hydrogen ions in water are bases.
- Acidic and basic solutions can conduct electricity.

pH

- pH measures how acidic or basic a solution is.
- The scale ranges from 0 to 14.

Neutralization

- Neutralization is the interaction between an acid and a base to form water and a salt.

Self Check

1. **Identify** what ions are produced by acids in water and bases in water. Give two properties each of acids and bases.

2. **Name** three acids and three bases and list an industrial or household use of each.

3. **Explain** how the concentration of hydronium ions and hydroxide ions are related to pH.

4. **Think Critically** In what ways might a company that uses a strong acid handle an acid spill on the factory floor?

Applying Math

5. **Solve One-Step Equations** How much more acidic is a solution with a pH of 2 than one with a pH of 6? How much more basic is a solution with a pH of 13 than one with a pH of 10?

Testing pH Using Natural Indicators

Goals

- **Determine** the relative acidity or basicity of several common solutions.
- **Compare** the strengths of several common acids and bases.

Materials

small test tubes (9)
test-tube rack
concentrated red cabbage
  juice in a dropper bottle
labeled bottles containing:
  household ammonia,
  baking soda solution,
  soap solution,
  0.1*M* hydrochloric acid
  solution, white vinegar,
  colorless carbonated
  soft drink, borax soap
  solution, distilled water
grease pencil
droppers (9)

Safety Precautions

WARNING: *Many acids and bases are poisonous, can damage your eyes, and can burn your skin. Wear goggles and gloves AT ALL TIMES. Tell your teacher immediately if a substance spills. Wash your hands after you finish but before removing your goggles.*

▶ Real-World Question

You have learned that certain substances, called indicators, change color when the pH of a solution changes. The juice from red cabbage is a natural indicator. How do the pH values of various solutions compare to each other? How can you use red cabbage juice to determine the relative pH of several solutions?

▶ Procedure

1. **Design** a data table to record the names of the solutions to be tested, the colors caused by the added cabbage juice indicator, and the relative strengths of the solutions.

2. Mark each test tube with the identity of the acid or base solution it will contain.

3. Half-fill each test tube with the solution to be tested.
   WARNING: *If you spill any liquids on your skin, rinse the area immediately with water. Alert your teacher if any liquid spills in the work area or on your skin.*

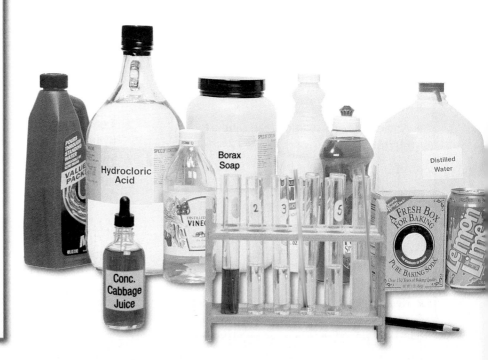

4. Add ten drops of the cabbage juice indicator to each of the solutions to be tested. Gently agitate or wiggle each test tube to mix the cabbage juice with the solution.

5. **Observe** and record the color of each solution in your data table.

▶ Analyze Your Data

1. **Compare** your observations with the table above. Record in your data table the relative acid or base strength of each solution you tested.

2. **List** the solutions by pH value starting with the most acidic and finishing with the most basic.

Determining pH Values	
Cabbage Juice Color	Relative Strength of Acid or Base
	strong acid
	medium acid
	weak acid
	neutral
	weak base
	medium base
	strong base

▶ Conclude and Apply

1. **Classify** which solutions were acidic and which were basic.

2. **Identify** which solution was the weakest acid. The strongest base? The closest to neutral?

3. **Predict** what ion might be involved in the cleaning process based upon your data for the ammonia, soap, and borax soap solutions.

▶ Form a Hypothesis

Form a hypothesis that explains why the borax soap solution was less basic than an ammonia solution of approximately the same concentration.

Communicating Your Data

Use your data to create labels for the solutions you tested. Include the relative strength of each solution and any other safety information you think is important on each label. **For more help, refer to the Science Skill Handbook.**

L ◆ 87

Salty Solutions

Did you know...

...Seawater is certainly a salty solution. Ninety-nine percent of all salt ions in the sea are sodium, chlorine, sulfate, magnesium, calcium, and potassium. The major gases in the sea are nitrogen, oxygen, carbon dioxide, argon, neon, and helium.

...Tears and saliva have a lot in common. Both are salty solutions that protect you from harmful bacteria, keep tissues moist, and help spread nutrients. Bland-tasting saliva, however, is 99 percent water. The remaining one percent is a combination of many ions, including sodium and several proteins.

...The largest salt lake in the United States is the Great Salt Lake. It covers more than 4,000 km² in Utah and is up to 13.4 m deep. The Great Salt Lake and the Salt Lake Desert were once part of the enormous, prehistoric Lake Bonneville, which was 305 m deep at some points.

Applying Math At its largest, Lake Bonneville covered about 32,000 km². What percentage of that area does the Great Salt Lake now cover?

...Salt can reduce pain. Gargled salt water is a disinfectant; it fights the bacteria that cause some sore throats.

Graph It

Visit bookl.msscience.com/science_stats **to research and learn about other elements in seawater. Create a graph that shows the amounts of the ten most common elements in 1 L of seawater.**

Reviewing Main Ideas

Section 1 What is a solution?

1. Elements and compounds are pure substances, because their compositions are fixed. Mixtures are not pure substances.

2. Heterogeneous mixtures are not mixed evenly. Homogeneous mixtures, also called solutions, are mixed evenly on a molecular level.

3. Solutes and solvents can be gases, liquids, or solids, combined in many different ways.

Section 2 Solubility

1. Because water molecules are polar, they can dissolve many different solutes. Like dissolves like.

2. Temperature and pressure can affect solubility.

3. Solutions can be unsaturated, saturated, or supersaturated, depending on how much solute is dissolved compared to the solubility of the solute in the solvent.

4. The concentration of a solution is the amount of solute in a particular volume of solvent.

Section 3 Acidic and Basic Solutions

1. Acids release H+ ions and produce hydronium ions when they are dissolved in water. Bases accept H+ ions and produce hydroxide ions when dissolved in water.

2. pH expresses the concentrations of hydronium ions and hydroxide ions in aqueous solutions.

3. In a neutralization reaction, an acid reacts with a base to form water and a salt.

Visualizing Main Ideas

Copy and complete the concept map on the classification of matter.

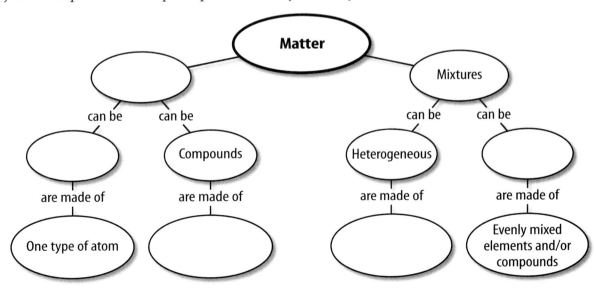

Using Vocabulary

acid p. 78
aqueous p. 70
base p. 81
concentration p. 75
heterogeneous
   mixture p. 65
homogeneous
   mixture p. 66
hydronium ion p. 78
indicator p. 84

neutralization p. 84
pH p. 82
precipitate p. 66
saturated p. 74
solubility p. 73
solute p. 66
solution p. 66
solvent p. 66
substance p. 64

Fill in the blanks with the correct vocabulary word.

1. A base has a(n) _____ value above 7.

2. A measure of how much solute is in a solution is its _____.

3. The amount of a solute that can dissolve in 100 g of solvent is its _____.

4. The _____ is the substance that is dissolved to form a solution.

5. The reaction between an acidic and basic solution is called _____.

6. A(n) _____ has a fixed composition.

Checking Concepts

Choose the word or phrase that best answers the question.

7. Which of the following is a solution?
   A) pure water
   B) an oatmeal-raisin cookie
   C) copper
   D) vinegar

8. What type of compounds will not dissolve in water?
   A) polar   **C)** nonpolar
   B) ionic   **D)** charged

9. What type of molecule is water?
   A) polar   **C)** nonpolar
   B) ionic   **D)** precipitate

10. When chlorine compounds are dissolved in pool water, what is the water?
    A) the alloy
    B) the solvent
    C) the solution
    D) the solute

11. A solid might become less soluble in a liquid when you decrease what?
    A) particle size   **C)** temperature
    B) pressure   **D)** container size

12. Which acid is used in the industrial process known as pickling?
    A) hydrochloric   **C)** sulfuric
    B) carbonic   **D)** nitric

13. A solution is prepared by adding 100 g of solid sodium hydroxide, NaOH, to 1,000 mL of water. What is the solid NaOH called?
    A) solution   **C)** solvent
    B) solute   **D)** mixture

14. Given equal concentrations, which of the following will produce the most hydronium ions in an aqueous solution?
    A) a strong base   **C)** a strong acid
    B) a weak base   **D)** a weak acid

15. Bile, an acidic body fluid used in digestion, has a high concentration of hydronium ions. Predict its pH.
    A) 11   **C)** less than 7
    B) 7   **D)** greater than 7

16. When you swallow an antacid, what happens to your stomach acid?
    A) It is more acidic.
    B) It is concentrated.
    C) It is diluted.
    D) It is neutralized.

Science Online bookl.msscience.com/vocabulary_puzzlemaker

Thinking Critically

17. **Infer** why deposits form in the steam vents of irons in some parts of the country.

18. **Explain** if it is possible to have a dilute solution of a strong acid.

19. **Draw Conclusions** Antifreeze is added to water in a car's radiator to prevent freezing in cold months. It also prevents overheating or boiling. Explain how antifreeze does both.

Use the illustration below to answer question 20.

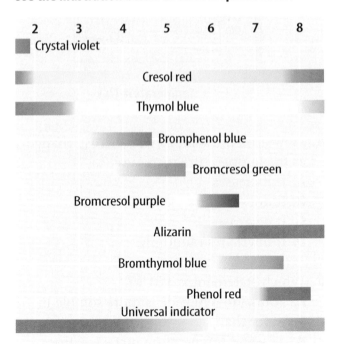

20. **Interpret** Chemists use a variety of indicators. Using the correct indicator is important. The color change must occur at the proper pH or the results could be misleading. Looking at the indicator chart, what indicators could be used to produce a color change at both pH 2 and pH 8?

21. **Explain** Water molecules can break apart to form H^+ ions and OH^- ions. Water is known as an amphoteric substance, which is something that can act as an acid or a base. Explain how this can be so.

22. **Describe** how a liquid-solid solution forms. How is this different from a liquid-gas solution? How are these two types of solutions different from a liquid-liquid solution? Give an example of each with your description.

23. **Compare and contrast** examples of heterogeneous and homogeneous mixtures from your daily life.

24. **Form a Hypotheses** A warm carbonated beverage seems to fizz more than a cold one when it is opened. Explain this based on the solubility of carbon dioxide in water.

Performance Activities

25. **Poem** Write a poem that explains the difference between a substance and a mixture.

Applying Math

Use the graph below to answer question 26.

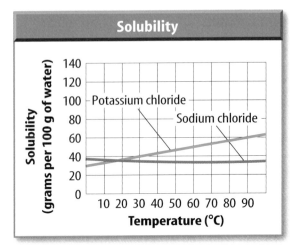

26. **Solubility** Using the solubility graph above, estimate the solubilities of potassium chloride and sodium chloride in grams per 100 g of water at 80°C.

27. **Juice Concentration** You made a one-liter (1,000 mL) container of juice. How much concentrate, in mL, did you add to make a concentration of 18 percent?

Part 1 | Multiple Choice

Record your answers on the answer sheet provided by your teacher or on a sheet of paper.

Use the illustration below to answer questions 1 and 2.

Composition of Earth's Atmosphere

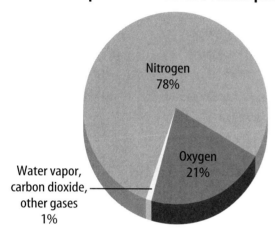

Nitrogen
78%

Oxygen
21%

Water vapor,
carbon dioxide,
other gases
1%

1. Which term best describes Earth's atmosphere?
   A. saturated
   B. solution
   C. precipitate
   D. indicator

2. Which of these is the solvent in Earth's atmosphere?
   A. nitrogen
   B. oxygen
   C. water vapor
   D. carbon dioxide

3. What characteristic do aqueous solutions share?
   A. They contain more than three solutes.
   B. No solids or gases are present as solutes in them.
   C. All are extremely concentrated.
   D. Water is the solvent in them.

Test-Taking Tip

Start the Day Right The morning of the test, eat a healthy breakfast with a balanced amount of protein and carbohydrates.

Use the illustration below to answer questions 4 and 5.

Solubility

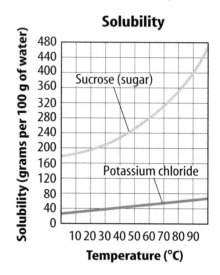

Sucrose (sugar)

Potassium chloride

Solubility (grams per 100 g of water)

Temperature (°C)

4. How does the solubility of sucrose change as the temperature increases?
   A. It increases.
   B. It does not change.
   C. It decreases.
   D. It fluctuates randomly.

5. Which statement is TRUE?
   A. Potassium chloride is more soluble in water than sucrose.
   B. As water temperature increases, the solubility of potassium chloride decreases.
   C. Sucrose is more soluble in water than potassium chloride.
   D. Water temperature has no effect on the solubility of these two chemicals.

6. Which of these is a property of acidic solutions?
   A. They taste sour.
   B. They feel slippery.
   C. They are in many cleaning products.
   D. They taste bitter.

Part 2 | Short Response/Grid In

Record your answers on the answer sheet provided by your teacher or on a sheet of paper.

7. Identify elements present in the alloy steel. Compare the flexibility and strength of steel and iron.

Use the illustration below to answer questions 8 and 9.

8. How can you tell that the matter in this bowl is a mixture?

9. What kind of mixture is this? Define this type of mixture, and give three additional examples.

10. Explain why a solute broken into small pieces will dissolve more quickly than the same type and amount of solute in large chunks.

11. Compare the concentration of two solutions: Solution A is composed of 5 grams of sodium chloride dissolved in 100 grams of water. Solution B is composed of 27 grams of sodium chloride dissolved in 100 grams of water.

12. Give the pH of the solutions vinegar, blood plasma, and ammonia. Compare the acidities of soft drinks, tomatoes, and milk.

13. Describe how litmus paper is used to determine the pH of a solution.

Part 3 | Open Ended

Record your answers on a sheet of paper.

14. Compare and contrast crystallization and a precipitation reaction.

15. Why is a carbonated beverage defined as a liquid-gas solution? In an open container, the ratio of liquid solvent to gas solute changes over time. Explain.

Use the illustration below to answer questions 16 and 17.

(Partial negative charge)

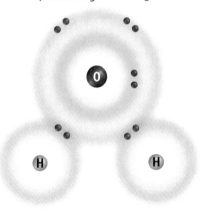

(Partial positive charge)

16. The diagram shows a water molecule. Use the distribution of electrons to describe this molecule's polarity.

17. Explain how the polarity of water molecules makes water effective in dissolving ionic compounds.

18. Marble statues and building facades in many of the world's cities weather more quickly today than when first constructed. Explain how the pH of water plays a role in this process.

19. Acetic acid, CH_3COOH, has more hydrogen atoms than the same concentration of hydrochloric acid, HCl. Hydrogen ions separate more easily from hydrochloric than acetic acid. Which acid is strongest? Why?

Carbon Chemistry

Better Camping Through Carbon Chemistry

When you are camping, you spend your nights sleeping under the stars. But one thing remains the same—carbon. Your food, clothing, body, and living things around you contain carbon compounds.

Science Journal Find and name four items around your classroom that are made from carbon compounds.

Start-Up Activities

Model Carbon's Bonding

Many of the compounds that compose your body and other living things are carbon compounds. This lab demonstrates some of the atomic combinations possible with one carbon and four other atoms. The ball represents a carbon atom. The toothpicks represent chemical bonds.

WARNING: *Do not eat any foods from this lab. Wash your hands before and after this lab.*

1. Insert four toothpicks into a small clay or plastic foam ball so they are evenly spaced around the sphere.

2. Make models of as many molecules as possible by adding raisins, grapes, and gumdrops to the ends of the toothpicks. Use raisins to represent hydrogen atoms, grapes to represent chlorine atoms, and gumdrops to represent fluorine atoms.

3. **Think Critically** Draw each model and write the formula for it in your Science Journal. What can you infer about the number of compounds a carbon atom can form?

Preview this chapter's content and activities at bookl.msscience.com

Hydrocarbons Make the following Foldable to help you learn the definitions of vocabulary words. This will help you understand the chapter content.

STEP 1 **Fold** a sheet of paper in half lengthwise. Make the back edge about 1.25 cm longer than the front edge.

STEP 2 **Fold** in half, then fold in half again to make three folds.

STEP 3 **Unfold and cut** only the top layer along the three folds to make four tabs.

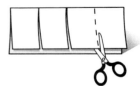

STEP 4 **Label** the tabs as shown.

Find Main Ideas As you read the chapter, find the definitions for each vocabulary word and write them under the appropriate tabs. Add additional words and definitions to help you understand your reading. List examples of each type of hydrocarbon under the appropriate tab.

Simple Organic Compounds

as you read

What You'll Learn

- **Explain** why carbon is able to form many compounds.
- **Describe** how saturated and unsaturated hydrocarbons differ.
- **Identify** isomers of organic compounds.

Why It's Important

Plants, animals, and many of the things that are part of your life are made of organic compounds.

🔍 Review Vocabulary

chemical bond: force that holds two atoms together

New Vocabulary

- organic compound
- hydrocarbon
- saturated hydrocarbon
- unsaturated hydrocarbon
- isomer

Organic Compounds

Earth's crust contains less than one percent carbon, yet all living things on Earth are made of carbon-containing compounds. Carbon's ability to bond easily and form compounds is the basis of life on Earth. A carbon atom has four electrons in its outer energy level, so it can form four covalent bonds with as many as four other atoms. When carbon atoms form four covalent bonds, they obtain the stability of a noble gas with eight electrons in their outer energy level. One of carbon's most frequent partners in forming covalent bonds is hydrogen.

Substances can be classified into two groups—those derived from living things and those derived from nonliving things, as shown in **Figure 1.** Most of the substances associated with living things contain carbon and hydrogen. These substances were called organic compounds, which means "derived from a living organism." However, in 1828, scientists discovered that living organisms are not necessary to form organic compounds. Despite this, scientists still use the term **organic compound** for most compounds that contain carbon.

✔ **Reading Check** *What is the origin of the term organic compound?*

Figure 1 Most substances can be classified as living or nonliving things.

Living things and products made from living things such as this wicker chair contain carbon.

Most of the things in this photo are nonliving and are composed of elements other than carbon.

Hydrocarbons

Many compounds are made of only carbon and hydrogen. A compound that contains only carbon and hydrogen atoms is called a **hydrocarbon.** The simplest hydrocarbon is methane, the primary component of natural gas. If you have a gas stove or gas furnace in your home, methane usually is the fuel that is burned in these appliances. Methane consists of a single carbon atom covalently bonded to four hydrogen atoms. The formula for methane is CH_4. **Figure 2** shows a model of the methane molecule and its structural formula. In a structural formula, the line between one atom and another atom represents a pair of electrons shared between the two atoms. This pair forms a single bond. Methane contains four single bonds.

Now, visualize the removal of one of the hydrogen atoms from a methane molecule, as in **Figure 3A.** A fragment of the molecule called a methyl group, $-CH_3$, would remain. The methyl group then can form a single bond with another methyl group. If two methyl groups bond with each other, the result is the two-carbon hydrocarbon ethane, C_2H_6, which is shown with its structural formula in **Figure 3B.**

Figure 2 Methane is the simplest hydrocarbon molecule. **Explain** why this is true.

Methane
CH_4

Figure 3 Here's a way to visualize how larger hydrocarbons are built up.

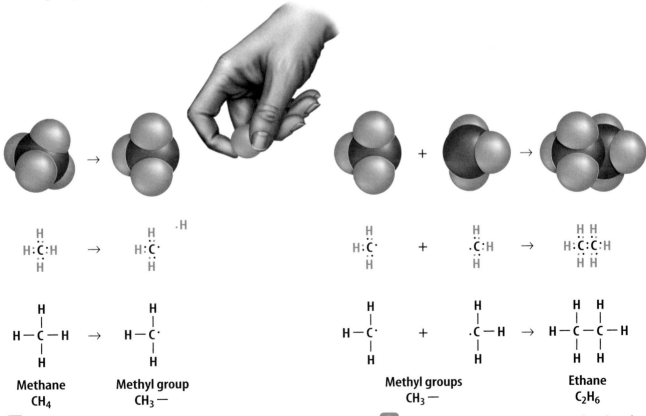

Methane
CH_4

Methyl group
$CH_3 -$

Methyl groups
$CH_3 -$

Ethane
C_2H_6

A A hydrogen is removed from a methane molecule, forming a methyl group.

B Each carbon atom in ethane has four bonds after the two methyl groups join.

Figure 4 Propane and butane are two useful fuels. **Explain** why they are called "saturated."

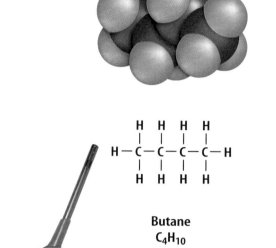

Propane
C₃H₈

When propane burns, it releases energy as the chemical bonds are broken. Propane often is used to fuel camp stoves and outdoor grills.

Butane
C₄H₁₀

Butane also releases energy when it burns. Butane is the fuel that is used in disposable lighters.

Hydrocarbons Petroleum is a mixture of hydrocarbons that formed from plants and animals that lived in seas and lakes hundreds of millions of years ago. With the right temperature and pressure, this plant and animal matter, buried deep under Earth's surface, decomposed to form petroleum. Why is petroleum a nonrenewable resource?

Saturated Hydrocarbons Methane and ethane are members of a series of molecules in which carbon and hydrogen atoms are joined by single covalent bonds. When all the bonds in a hydrocarbon are single bonds, the molecule is called a **saturated hydrocarbon.** It is called *saturated* because no additional hydrogen atoms can be added to the molecule. The carbon atoms are saturated with hydrogen atoms. The formation of larger hydrocarbons occurs in a way similar to the formation of ethane. A hydrogen atom is removed from ethane and replaced by a $-CH_3$ group. Propane, with three carbon atoms, is the third member of the series of saturated hydrocarbons. Butane has four carbon atoms. Both of these hydrocarbons are shown in **Figure 4.** The names and the chemical formulas of a few of the smaller saturated hydrocarbons are listed in **Table 1.** Saturated hydrocarbons are named with an *-ane* ending. Another name for these hydrocarbons is alkanes.

Reading Check *What is a saturated hydrocarbon?*

These short hydrocarbon chains have low boiling points, so they evaporate and burn easily. That makes methane a good fuel for your stove or furnace. Propane is used in gas grills, lanterns, and to heat air in hot-air balloons. Butane often is used as a fuel for camp stoves and lighters. Longer hydrocarbons are found in oils and waxes. Carbon can form long chains that contain hundreds or even thousands of carbon atoms. These extremely long chains make up many of the plastics that you use.

Unsaturated Hydrocarbons Carbon also forms hydrocarbons with double and triple bonds. In a double bond, two pairs of electrons are shared between two atoms, and in a triple bond, three pairs of electrons are shared. Hydrocarbons with double or triple bonds are called **unsaturated hydrocarbons.** This is because the carbon atoms are not saturated with hydrogen atoms.

Ethene, the simplest unsaturated hydrocarbon, has two carbon atoms joined by a double bond. Propene is an unsaturated hydrocarbon with three carbons. Some unsaturated hydrocarbons have more than one double bond. Butadiene (byew tuh DI een) has four carbon atoms and two double bonds. The structures of ethene, propene, and butadiene are shown in **Figure 5.**

Unsaturated compounds with at least one double bond are named with an -*ene* ending. Notice that the names of the compounds below have an -*ene* ending. These compounds are called alkenes.

Table 1 The Structures of Hydrocarbons		
Name	**Structural Formula**	**Chemical Formula**
Methane	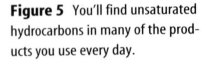	CH_4
Ethane		C_2H_6
Propane		C_3H_8
Butane		C_4H_{10}

Reading Check *What type of bonds are found in unsaturated hydrocarbons?*

Figure 5 You'll find unsaturated hydrocarbons in many of the products you use every day.

Ethene

Ethene helps ripen fruits and vegetables. It's also used to make milk and soft-drink bottles.

Propene

This detergent bottle contains the tough plastic polypropylene, which is made from propene.

Butadiene

Butadiene made it possible to replace natural rubber with synthetic rubber.

$$H-C\equiv C-H$$

Ethyne or Acetylene
C_2H_2

Modeling Isomers

Procedure

WARNING: *Do not eat any foods in this lab.*

1. Construct a model of pentane, C_5H_{12}. Use **toothpicks** for covalent bonds and small balls of different **colored clay or gumdrops** for carbon and hydrogen atoms.
2. Using the same materials, build a molecule with a different arrangement of the atoms. Are there any other possibilities?
3. Make a model of hexane, C_6H_{14}.
4. Arrange the atoms of hexane in different ways.

Analysis

1. How many isomers of pentane did you build? How many isomers of hexane?
2. Do you think there are more isomers of heptane, C_7H_{16}, than hexane? Explain.

Try at Home

Triple Bonds Unsaturated hydrocarbons also can have triple bonds, as in the structure of ethyne (EH thine) shown in **Figure 6.** Ethyne, commonly called acetylene (uh SE tuh leen), is a gas used for welding because it produces high heat as it burns. Welding torches mix acetylene and oxygen before burning. These unsaturated compounds are called alkynes.

Hydrocarbon Isomers Suppose you had ten blocks that could be snapped together in different arrangements. Each arrangement of the same ten blocks is different. The atoms in an organic molecule also can have different arrangements but still have the same molecular formula. Compounds that have the same molecular formula but different arrangements, or structures, are called **isomers** (I suh murz). Two isomers, butane and isobutane, are shown in **Figure 7.** They have different chemical and physical properties because of their different structures. As the size of a hydrocarbon molecule increases, the number of possible isomers also increases.

By now, you might be confused about how organic compounds are named. **Figure 8** explains the system that is used to name simple organic compounds.

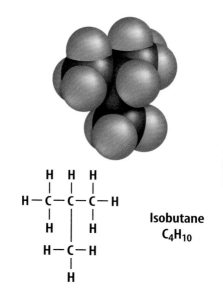

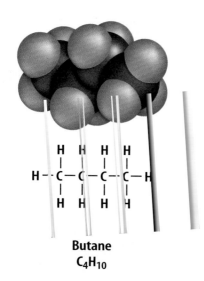

Isobutane
C_4H_{10}

Butane
C_4H_{10}

Figure 7 Butane and isobutane both have four carbons and ten hydrogens but their structures and properties are different.

Figure 8

More than one million organic compounds have been discovered and created, and thousands of new ones are synthesized in laboratories every year. To keep track of all these carbon-containing molecules, the International Union of Pure and Applied Chemistry, or IUPAC, devised a special naming system (a nomenclature) for organic compounds. As shown here, different parts of an organic compound's name—its root, suffix, or prefix—give information about its size and structure.

Carbon atoms	Name	Molecular formula
1	Methane	CH_4
2	Ethane	CH_3CH_3
3	Propane	$CH_3CH_2CH_3$
4	Butane	$CH_3CH_2CH_2CH_3$
5	Pentane	$CH_3CH_2CH_2CH_2CH_3$
6	Hexane	$CH_3CH_2CH_2CH_2CH_2CH_3$
7	Heptane	$CH_3CH_2CH_2CH_2CH_2CH_2CH_3$
8	Octane	$CH_3CH_2CH_2CH_2CH_2CH_2CH_2CH_3$
9	Nonane	$CH_3CH_2CH_2CH_2CH_2CH_2CH_2CH_2CH_3$
10	Decane	$CH_3CH_2CH_2CH_2CH_2CH_2CH_2CH_2CH_2CH_3$

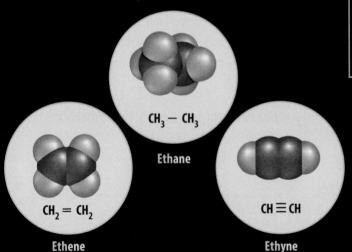

$CH_3 - CH_3$

Ethane

$CH_2 = CH_2$

Ethene

$CH \equiv CH$

Ethyne

ROOT WORDS The key to every name given to a compound in organic chemistry is its root word. This word tells how many carbon atoms are found in the longest continuous carbon chain in the compound. Except for compounds with one to four carbon atoms, the root word is based on Greek numbers.

SUFFIXES The suffix of the name for an organic compound indicates the kind of covalent bonds joining the compound's carbon atoms. If the atoms are joined by single covalent bonds, the compound's name will end in **-ane**. If there is a double covalent bond in the carbon chain, the compound's name ends in **-ene**. Similarly, if there is a triple bond in the chain, the compound's name will end in **-yne**.

PREFIXES The prefix of the name for an organic compound describes how the carbon atoms in the compound are arranged. Organic molecules that have names with the prefix **cyclo-** contain a ring of carbon atoms. For example, cyclopentane contains five carbon atoms all joined by single bonds in a ring.

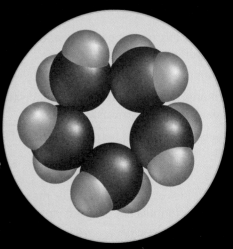

Cyclopentane

Figure 9 Visualize a hydrogen atom removed from a carbon atom on both ends of a hexane chain. The two end carbons form a bond with each other. **Describe** *how the chemical formula changes.*

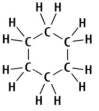

Hexane
C_6H_{14}

Cyclohexane
C_6H_{12}

Hydrocarbons in Rings You might be thinking that all hydrocarbons are chains of carbon atoms with two ends. Some molecules contain rings. You can see the structures of two different molecules in **Figure 9.** The carbon atoms of hexane bond together to form a closed ring containing six carbons. Each carbon atom still has four bonds. The prefix *cyclo-* in their names tells you that the molecules are cyclic, or ring shaped.

Ring structures are not uncommon in chemical compounds. Many natural substances such as sucrose, glucose, and fructose are ring structures. Ring structures can contain one or more double bonds.

 **Reading Check** *What does the prefix* cyclo- *tell you about a molecule?*

---

section 1 review

Summary

Organic Compounds

- All living things contain carbon.
- Carbon atoms form covalent bonds.

Hydrocarbons

- Hydrocarbons are compounds that contain only hydrogen and carbon.
- The simplest hydrocarbon is methane.
- Saturated hydrocarbons are compounds that contain only single covalent bonds.
- Unsaturated hydrocarbons are compounds that form double and triple bonds.

Isomers

- Isomers are compounds that have the same molecular formula but different structures.
- Isomers also have different chemical and physical properties.

Self Check

1. **Describe** a carbon atom.
2. **Identify** Give one example of each of the following: a compound with a single bond, a compound with a double bond, and a compound with a triple bond. Write the chemical formula and draw the structure for each.
3. **Draw** all the possible isomers for heptane, C_7H_{16}.
4. **Think Critically** Are propane and cylcopropane isomers? Draw their structures. Use the structures and formulas to explain your answer.

Applying Math

5. **Make and Use Graphs** From **Table 1,** plot the number of carbon atoms on the *x*-axis and the number of hydrogen atoms on the *y*-axis. Predict the formula for the saturated hydrocarbon that has 11 carbon atoms.

Other Organic Compounds

Substituted Hydrocarbons

Suppose you pack an apple in your lunch every day. One day, you have no apples, so you substitute a pear. When you eat your lunch, you'll notice a difference in the taste and texture of your fruit. Chemists make substitutions, too. They change hydrocarbons to make compounds called substituted hydrocarbons. To make a substituted hydrocarbon, one or more hydrogen atoms are replaced by atoms such as halogens or by groups of atoms. Such changes result in compounds with chemical properties different from the original hydrocarbon. When one or more chlorine atoms are added to methane in place of hydrogens, new compounds are formed. **Figure 10** shows the four possible compounds formed by substituting chlorine atoms for hydrogen atoms in methane.

as you read

What You'll Learn
- **Describe** how new compounds are formed by substituting hydrogens in hydrocarbons.
- **Identify** the classes of compounds that result from substitution.

Why It's Important
Many natural and manufactured organic compounds are formed by replacing hydrogen with other atoms.

 Review Vocabulary
chemical formula: chemical shorthand that uses symbols to tell what elements are in a compound and their ratios

New Vocabulary
- hydroxyl group
- carboxyl group
- amino group
- amino acid

Chloromethane contains a single chlorine atom.

$$H - \overset{\overset{\textstyle H}{|}}{\underset{\underset{\textstyle H}{|}}{C}} - Cl$$

CH_3Cl

Trichloromethane, or chloroform, has three chlorine atoms. It is used in the production of fluoropolymers—one of the raw materials used to make nonstick coating.

$$H - \overset{\overset{\textstyle Cl}{|}}{\underset{\underset{\textstyle Cl}{|}}{C}} - Cl$$

$CHCl_3$

Dichloromethane contains two chlorine atoms. This is used in some paint and varnish removers.

$$H - \overset{\overset{\textstyle Cl}{|}}{\underset{\underset{\textstyle H}{|}}{C}} - Cl$$

CH_2Cl_2

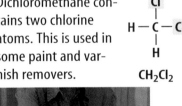

Carbon tetrachloride is a fully substituted methane molecule with four chlorines.

$$Cl - \overset{\overset{\textstyle Cl}{|}}{\underset{\underset{\textstyle Cl}{|}}{C}} - Cl$$

CCl_4

Figure 10 Chlorine can replace hydrogen atoms in methane.

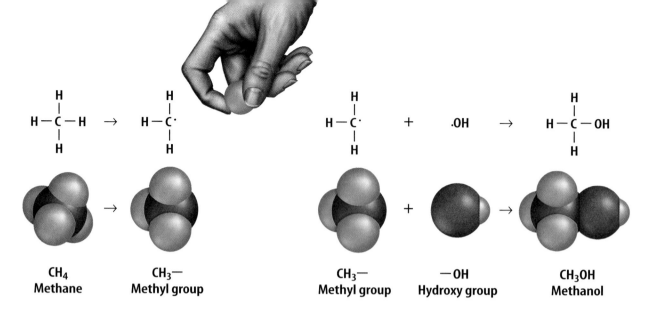

H—C—H → H—C·

CH₄ → CH₃—
Methane   Methyl group

H—C· + .OH → H—C—OH

CH₃— + —OH → CH₃OH
Methyl group   Hydroxy group   Methanol

Figure 11 After the methane molecule loses one of its hydrogen atoms, it has an extra electron to share, as does the hydroxyl group. **Identify** *the type of bond formed.*

Alcohols Groups of atoms also can be added to hydrocarbons to make different compounds. The **hydroxyl** (hi DROK sul) **group** is made up of an oxygen atom and a hydrogen atom joined by a covalent bond. A hydroxyl group is represented by the formula −OH. An alcohol is formed when a hydroxyl group replaces a hydrogen atom in a hydrocarbon. **Figure 11** shows the formation of the alcohol methanol. A hydrogen atom in the methane molecule is replaced by a hydroxyl group.

Reading Check *What does the formula* −OH *represent?*

Larger alcohol molecules are formed by adding more carbon atoms to the chain. Ethanol is an alcohol produced naturally when sugar in corn, grains, and fruit ferments. It is a combination of ethane, which contains two carbon atoms, and an −OH group. Its formula is C_2H_5OH. Isopropyl alcohol forms when the hydroxyl group is substituted for a hydrogen atom on the middle carbon of propane rather than one of the end carbons. **Table 2** lists three alcohols with their structures and uses. You've probably used isopropyl alcohol to disinfect injuries. Did you know that ethanol can be added to gasoline and used as a fuel for your car?

Table 2 Common Alcohols

Uses	Methanol	Ethanol	Isopropyl Alcohol
Fuel	yes	yes	no
Cleaner	yes	yes	yes
Disinfectant	no	yes	yes
Manufacturing chemicals	yes	yes	yes

Carboxylic Acids Have you ever tasted vinegar? Vinegar is a solution of acetic acid and water. You can think of acetic acid as the hydrocarbon methane with a carboxyl (car BOK sul) group substituted for a hydrogen. A **carboxyl group** consists of a carbon atom that has a double bond with one oxygen atom and a single bond with a hydroxyl group. Its formula is $-COOH$. When a carboxyl group is substituted in a hydrocarbon, the substance formed is called a carboxylic acid. The simplest carboxylic acid is formic acid. Formic acid consists of a single hydrogen atom and a carboxyl group. You can see the structures of formic acid and acetic acid in **Figure 12.**

You probably can guess that many other carboxylic acids are formed from longer hydrocarbons. Many carboxylic acids occur in foods. Citric acid is found in citrus fruits such as oranges and grapefruit. Lactic acid is present in sour milk. Acetic acid dissolved in water—vinegar—often is used in salad dressings.

Methanoic, or
formic, acid
HCOOH

Ethanoic, or
acetic, acid
CH₃COOH

Figure 12 *Crematogaster* ants make the simplest carboxylic acid, formic acid. Notice the structure of the $-COOH$ group.
Describe *how the structures of formic acid and acetic acid differ.*

Amines Substituted hydrocarbons, called amines, formed when an amino (uh ME noh) group replaces a hydrogen atom. An **amino group** is a nitrogen atom joined by covalent bonds to two hydrogen atoms. It has the formula $-NH_2$. Methylamine, shown in **Figure 13,** is formed when one of the hydrogens in methane is replaced with an amino group. A more complex amine is the novocaine dentists once used to numb your mouth during dental work. Amino groups are important because they are a part of many biological compounds that are essential for life. When an amino group bonds with one additional hydrogen atom, the result is ammonia, NH_3.

Figure 13 Complex amines account for the strong smells of cheeses such as these.

Amino Acids You have seen that a carbon group can be substituted onto one end of a chain to make a new molecule. It's also possible to substitute groups on both ends of a chain and even to replace hydrogen atoms bonded to carbon atoms in the middle of a chain. When both an amino group ($-NH_2$) and a carboxyl acid group ($-COOH$) replace hydrogens on the same carbon atom in a molecule, a type of compound known as an **amino acid** is formed. Amino acids are essential for human life.

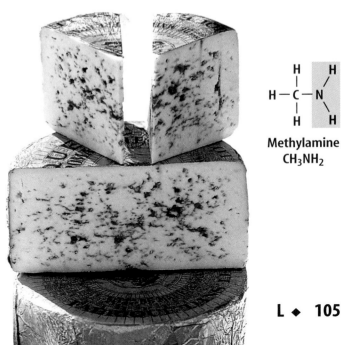

Methylamine
CH₃NH₂

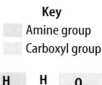

Key
Amine group
Carboxyl group

H
|
H   H   O
| | //
N — C — C
| | \
H   H   O — H

Glycine

H
|
H — C — H
|
H | O
| | //
N — C — C
| | \
H   H   O — H

Alanine

Figure 14 The 20 amino acids needed for protein synthesis each contain a central carbon atom bonded to an amine group, a hydrogen atom, and a carboxyl group. The fourth bond, shown in yellow, is different for each amino acid.

The Building Blocks of Protein

Amino acids are the building blocks of proteins, which are an important class of biological molecules needed by living cells. Twenty different amino acids bond in different combinations to form the variety of proteins that are needed in the human body. Glycine and alanine are shown in **Figure 14.**

Glycine is the simplest amino acid. It is a methane molecule in which one hydrogen atom has been replaced by an amine group and another has been replaced by a carboxyl group. The other 19 amino acids are formed by replacing the yellow highlighted hydrogen atom with different groups. For example, in alanine, one hydrogen atom is replaced by a methyl $(-CH_3)$ group.

✔ **Reading Check** *What are the building blocks of protein?*

Some amino acids, such as glycine and alanine, are manufactured within the human body. They are called nonessential amino acids. This means that it is not essential to consume these types of amino acids. More than half of the twenty amino acids are considered nonessential. The essential amino acids, those that must be consumed, are obtained by eating protein-rich foods. These foods include meat, eggs, and milk.

section 2 review

Summary

Substituted Hydrocarbons

- A substituted hydrocarbon has one or more hydrogen atoms replaced.
- The chemical properties of the substituted hydrocarbon are different from the original hydrocarbon.

Types of Substitutions

- Alcohols are made when a hydroxyl group is substituted for a hydrogen atom.
- Carboxylic acids are formed when the carboxyl group is substituted for a hydrogen atom.
- When an amino group is substituted for hydrogen, an amine is formed.
- Amino acids have both an amino group and a carboxyl group.
- Twenty amino acids are building blocks of proteins needed in the human body.

Self Check

1. **Draw** Tetrafluoroethylene is a substituted hydrocarbon in which all four of the hydrogen atoms are replaced by fluorine. Draw the structural formula for this molecule.

2. **Describe** how the 20 amino acids differ from each other.

3. **Identify** Starting with a hexane molecule, C_6H_{14}, draw and label each new molecule when adding an alcohol group, a carboxylic group, and an amino group.

4. **Think Critically** The formula for one compound that produces the odor in skunk spray is $CH_3CH_2CH_2CH_2SH$. Draw and examine the structural formula. Does it fit the definition of a substituted hydrocarbon? Explain.

Applying Skills

5. **Define** Compounds in which hydrogen atoms have been replaced by chlorine and fluorine atoms are known as chlorofluorocarbons (CFCs). Draw the structures of the four CFCs using CH_4 as the starting point.

Science Online   bookl.msscience.com/self_check_quiz

Conversion of Alcohols

Have you ever wondered how chemists change one substance into another? In this lab, you will change an alcohol into an acid.

▶ Real-World Question

What changes occur when ethanol is exposed to conditions like those produced by exposure to air and bacteria?

Goals

■ **Observe** a chemical change in an alcohol.
■ **Infer** the product of the chemical change.

Materials

test tube and stopper
test-tube rack
pH test paper
10-mL graduated cylinders (2)
dropper
0.01M potassium permanganate solution (1 mL)
6.0M sodium hydroxide solution (1 mL)
ethanol (3 drops)

Safety Precautions

WARNING: *Handle these chemicals with care. Immediately flush any spills with water and call your teacher. Keep your hands away from your face.*

▶ Procedure

1. Measure 1 mL of potassium permanganate solution and pour it into a test tube. Carefully measure 1 mL of sodium hydroxide solution and add it to the test tube.

2. With your teacher's help, dip a piece of pH paper into the mixture in the test tube. Record the result in your Science Journal.

Alcohol Conversion

Procedure Step	Observations
Step 2	
Step 3	Do not write in
Step 4	this book.
Step 5	

3. Add three drops of ethanol to the test tube. Put a stopper on the test tube and gently shake it for one minute. Record any changes.

4. Place the test tube in a test-tube rack. Observe and record any changes you notice during the next five minutes.

5. Test the sample with pH paper again. Record what you observe.

6. Your teacher will dispose of the solutions.

▶ Conclude and Apply

1. **Analyze Results** Did a chemical reaction take place? What leads you to infer this?

2. **Predict** Alcohols can undergo a chemical reaction to form carboxylic acids in the presence of potassium permanganate. If the alcohol used is ethanol, what would you predict to be the chemical formula of the acid produced?

𝒞ommunicating

Your Data

Compare your conclusions with other students in your class. **For more help, refer to the** Science Skill Handbook.

Biological Compounds

as you read

What You'll Learn

- **Describe** how large organic molecules are made.
- **Explain** the roles of organic molecules in the body.
- **Explain** why eating a balanced diet is important for good health.

Why It's Important

Polymers are organic molecules that are important to your body processes and everyday living.

⊙ **Review Vocabulary**

chemical reaction: process that produces chemical change, resulting in new substances that have properties different from those of the original substances

New Vocabulary

- polymer
- monomer
- polymerization
- protein
- carbohydrate
- sugars
- starches
- lipids
- cholesterol

What's a polymer?

Now that you know about some simple organic molecules, you can begin to learn about more complex molecules. One type of complex molecule is called a polymer (PAH luh mur). A **polymer** is a molecule made up of many small organic molecules linked together with covalent bonds to form a long chain. The small, organic molecules that link together to form polymers are called **monomers.** Polymers can be produced by living organisms or can be made in a laboratory. Polymers produced by living organisms are called natural polymers. Polymers made in a laboratory are called synthetic polymers.

✔ **Reading Check** *What is a polymer, and how does it resemble a chain?*

To picture what polymers are, it is helpful to start with small synthetic polymers. You use such polymers every day. Plastics, synthetic fabrics, and nonstick surfaces on cookware are polymers. The unsaturated hydrocarbon ethylene, C_2H_4, is the monomer of a common polymer used often in plastic bags. The monomers are bonded together in a chemical reaction called **polymerization** (puh lih muh ruh ZAY shun). As you can see in **Figure 15,** the double bond breaks in each ethylene molecule. The two carbon atoms then form new bonds with carbon atoms in other ethylene molecules. This process is repeated many times and results in a much larger molecule called polyethylene. A polyethylene molecule can contain 10,000 ethylene units.

Figure 15 Small molecules called monomers link into long chains to form polymers.

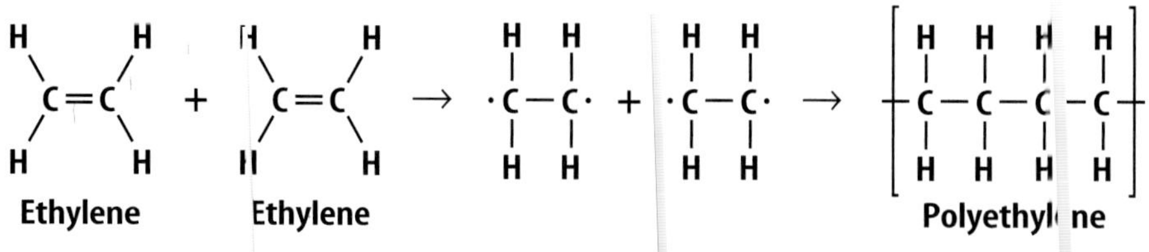

Ethylene   Ethylene   Polyethylene

The carbon atoms that were joined by the double bond each have an electron to share with another carbon in another molecule of ethylene. The process goes on until a long molecule is formed.

Glycine   Alanine

Proteins are Polymers

You've probably heard about proteins when you've been urged to eat healthful foods. A **protein** is a polymer that consists of a chain of individual amino acids linked together. Your body cannot function properly without them. Proteins in the form of enzymes serve as catalysts and speed up chemical reactions in cells. Some proteins make up the structural materials in ligaments, tendons, muscles, cartilage, hair, and fingernails. Hemoglobin, which carries oxygen through the blood, is a protein polymer, and all body cells contain proteins.

The various functions in your body are performed by different proteins. Your body makes many of these proteins by assembling 20 amino acids in different ways. Nine of the amino acids that are needed to make proteins cannot be produced by your body. These amino acids, which are called essential amino acids, must come from the food you eat. That's why you need to eat a diet containing protein-rich foods, like those in **Table 3.**

The process by which your body converts amino acids to proteins is shown in **Figure 16.** In this reaction, the amino group of the amino acid alanine forms a bond with the carboxyl group of the amino acid glycine, and a molecule of water is released. Each end of this new molecule can form similar bonds with another amino acid. The process continues in this way until the amino acid chain, or protein, is complete.

✔ **Reading Check** *How is an amino acid converted to protein?*

Table 3 Protein Content (Approximate)	
Foods	**Protein Content (g)**
Chicken breast (113 g)	28
Eggs (2)	12
Whole milk (240 mL)	8
Peanut butter (30 g)	8
Kidney beans (127 g)	8

Figure 16 Both ends of an amino acid can link with other amino acids.
Identify *the molecule that is released in the process.*

Mini LAB

Summing Up Protein

Procedure
1. Make a list of the foods you ate during the last 24 h.
2. Use the data your teacher gives you to find the total number of grams of protein in your diet for the day. Multiply the grams of protein in one serving of food by the number of units of food you ate. The recommended daily allowance (RDA) of protein for girls 11 to 14 years old is 46 g per day. For boys 11 to 14 years old, the RDA is 45 g per day.

Analysis
1. Was your total greater or less than the RDA?
2. Which of the foods you ate supplied the largest amount of protein? What percent of the total grams of protein did that food supply?

Figure 17 These foods contain a high concentration of carbohydrates.

Carbohydrates

The day before a race, athletes often eat large amounts of foods containing carbohydrates like the ones in **Figure 17.** What's in pasta and other foods like bread and fruit that gives the body a lot of energy? These foods contain sugars and starches, which are members of the family of organic compounds called carbohydrates. A **carbohydrate** is an organic compound that contains only carbon, hydrogen, and oxygen, usually in a ratio of two hydrogen atoms to one oxygen atom. In the body, carbohydrates are broken down into simple sugars that the body can use for energy. The different types of carbohydrates are divided into groups—sugars, starches, and cellulose.

Table 4 below gives some approximate carbohydrate content for some of the common foods.

Table 4 Carbohydrates in Foods (Approximate)	
Foods	Carbohydrate Content (g)
Apple (1)	21
White rice ($\frac{1}{2}$ cup)	17
Baked potato ($\frac{1}{2}$ cup)	15
Wheat bread (1 slice)	13
Milk (240 mL)	12

Figure 18 Glucose and fructose are simple six-carbon carbohydrates found in many fresh and packaged foods. Glucose and fructose are isomers.
Explain *why they are isomers.*

Glucose

Fructose

Sugars

If you like chocolate-chip cookies or ice cream, then you're familiar with sugars. They are the substances that make fresh fruits and candy sweet. Simple **sugars** are carbohydrates containing five, six, or seven carbon atoms arranged in a ring. The structures of glucose and fructose, two common simple sugars, are shown in **Figure 18.** Glucose forms a six-carbon ring. It is found in many naturally sweet foods, such as grapes and bananas. Fructose is the sweet substance found in ripe fruit and honey. It often is found in corn syrup and added to many foods as a sweetener. The sugar you probably have in your sugar bowl or use in baking a cake is sucrose. Sucrose, shown in **Figure 19,** is a combination of the two simple sugars glucose and fructose. In the body, sucrose cannot move through cell membranes. It must be broken down into glucose and fructose to enter cells. Inside the cells, these simple sugars are broken down further, releasing energy for cell functions.

Starches

Starches are large carbohydrates that exist naturally in grains such as rice, wheat, corn, potatoes, lima beans, and peas. **Starches** are polymers of glucose monomers in which hundreds or even thousands of glucose molecules are joined together. Because each sugar molecule releases energy when it is broken down, starches are sources of large amounts of energy.

Figure 19 Sucrose is a molecule of glucose combined with a molecule of fructose.
Identify *What small molecule must be added to sucrose when it separates to form the two six-carbon sugars?*

Sucrose

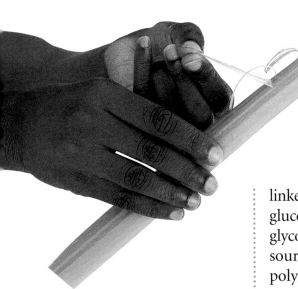

Figure 20 Your body cannot chemically break down the long cellulose fibers in celery, but it needs fiber to function properly.

Other Glucose Polymers Two other important polymers that are made up of glucose molecules are cellulose and glycogen. Cellulose makes up the long, stiff fibers found in the walls of plant cells, like the strands that pull off the celery stalk in **Figure 20.** It is a polymer that consists of long chains of glucose units linked together. Glycogen is a polymer that also contains chains of glucose units, but the chains are highly branched. Animals make glycogen and store it mainly in their muscles and liver as a ready source of glucose. Although starch, cellulose, and glycogen are polymers of glucose, humans can't use cellulose as a source of energy. The human digestive system can't convert cellulose into sugars. Grazing animals, such as cows, have special digestive systems that allow them to break down cellulose into sugars.

Reading Check *How do the location and structure of glycogen and cellulose differ?*

Applying Science

Which foods are best for quick energy?

Foods high in carbohydrates are sources of energy.

Identifying the Problem
The chart shows some foods and their carbohydrate count. Look at the differences in how much energy they might provide, given their carbohydrate count.

Solving the Problem
1. Create a high-energy meal with the most carbohydrates. Include one choice from each category. Create another meal that contains a maximum of 60 g of carbohydrates.
2. Meat and many vegetables have only trace amounts of carbohydrates. How many grams of carbohydrates would a meal of turkey, stuffing, lettuce salad, and lemonade contain?

Carbohydrate Counts for Common Foods					
Main Dish		**Side Dish**		**Drink**	
two slices wheat bread	26 g	fudge brownie	25 g	orange juice	27 g
macaroni and cheese	29 g	apple	21 g	cola	38 g
two pancakes	28 g	baked beans	27 g	sweetened iced tea	22 g
chicken and noodles	39 g	blueberry muffin	27 g	lemon-lime soda	38 g
hamburger with bun	34 g	cooked carrots	8 g	hot cocoa	25 g
hot oatmeal	25 g	banana	28 g	apple juice	29 g
plain bagel	38 g	baked potato	34 g	lemonade	28 g
bran flakes with raisins	47 g	stuffing	22 g	whole milk	12 g
lasagna	50 g	brown rice	22 g	chocolate milk	26 g
spaghetti with marinara	50 g	corn on the cob	14 g	sports drink	24 g

Lipids

A **lipid** is an organic compound that contains the same elements as carbohydrates—carbon, hydrogen, and oxygen—but in different proportions. They are the reaction products of glycerol, which has three –OH groups and three long-chain carboxylic acids, as pictured in **Figure 21.** Lipids are in many of the foods you eat such as the ones shown in **Figure 22.** Lipids are commonly called fats and oils, but they also are found in greases and waxes such as beeswax. Wax is a lipid, but it is harder than fat because of its chemical composition. Bees secrete wax from a gland in the abdomen to form beeswax, which is part of the honeycomb.

Lipids Store Energy Lipids store energy in their bonds, just as carbohydrates do, but they are a more concentrated source of energy than carbohydrates. If you eat more food than your body needs to supply you with the energy for usual activities, the excess energy from the food is stored by producing lipids.

How can energy be stored in a molecule? The chemical reaction that produces lipids is endothermic. An endothermic reaction is one in which energy is absorbed. This means that energy is stored in the chemical bonds of lipids. When your body needs energy, the bonds are broken and energy is released. This process protects your body when you need extra energy or when you might not be able to eat. If you regularly eat more food than you need, large amounts of lipids will be produced and stored as fat on your body.

✔ Reading Check *What is a lipid and how does your body use lipids to store energy?*

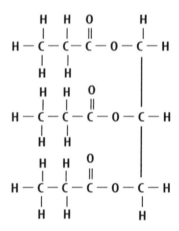

Figure 21 Lipids consist of two parts—glycerol and three molecules of carboxylic acid. **Identify** *which portion is from glycerol and which portion is from carboxylic acid.*

Figure 22 Many of the foods that you eat contain fats and oils, which are lipids.

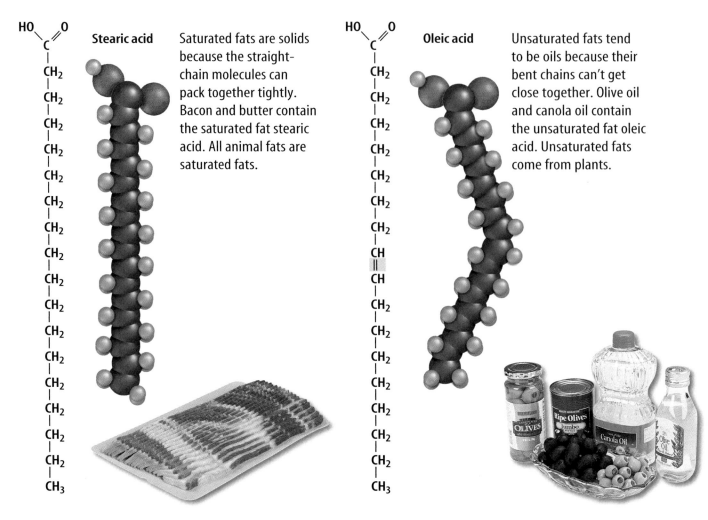

Stearic acid Saturated fats are solids because the straight-chain molecules can pack together tightly. Bacon and butter contain the saturated fat stearic acid. All animal fats are saturated fats.

Oleic acid Unsaturated fats tend to be oils because their bent chains can't get close together. Olive oil and canola oil contain the unsaturated fat oleic acid. Unsaturated fats come from plants.

Figure 23 Whether a lipid is a liquid or a solid depends on the type of bonds it has.

Saturated and Unsaturated Lipids Not all lipids are the same. Recall the difference between saturated and unsaturated hydrocarbons. Unsaturated molecules have one or more double or triple bonds between carbon atoms. Lipid molecules also can be saturated or unsaturated. As you can see in **Figure 23,** when a lipid is saturated, the acid chains are straight because all the bonds are single bonds. They are able to pack together closely. A compact arrangement of the molecules is typical of a solid such as margarine or shortening. These solid lipids consist mainly of saturated fats.

When a lipid is unsaturated, as in **Figure 23,** the molecule bends wherever there is a double bond. This prevents the chains from packing close together, so these lipids tend to be liquid oils such as olive or corn oil.

Doctors have observed that people who eat a diet high in saturated fats have an increased risk of developing cardiovascular problems such as heart disease. The effect of saturated fat seems to be increased blood cholesterol, which may be involved in the formation of deposits on artery walls. Fortunately, many foods containing unsaturated fats are available. Making wise choices in the foods that you eat can help keep your body healthy.

Cholesterol

Cholesterol is a complex lipid that is present in foods that come from animals, such as meat, butter, eggs, and cheese. However, cholesterol is not a fat. Even if you don't eat foods containing cholesterol, your body makes its own supply. Your body needs cholesterol for building cell membranes. Cholesterol is not found in plants, so oils derived from plants are free of cholesterol. However, the body can convert fats in these oils to cholesterol.

Deposits of cholesterol, called plaque, can build up on the inside walls of arteries. This condition, known as atherosclerosis, is shown in **Figure 24.** When arteries become clogged, the flow of blood is restricted, which results in high blood pressure. This, in turn, can lead to heart disease. Although the cause of plaque build up on the inside walls of arteries is unknown, limiting the amount of saturated fat and cholesterol might help to lower cholesterol levels in the blood and might help reduce the risk of heart problems.

Heart disease is a major health concern in the United States. As a result, many people are on low-cholesterol diets. What types of foods should people choose to lower their cholesterol level?

 **Reading Check** *What is atherosclerosis and why is this condition dangerous?*

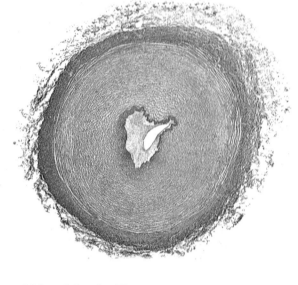

Figure 24 This view of an artery shows atherosclerosis, a dangerous condition in which arteries in the body become clogged. Deposits build up on the inside walls of the artery, leaving less room for blood to flow.

section 3 review

Summary

Polymers and Proteins

- A polymer is a molecule made up of small, repeating units.
- The small molecules that link together to form a polymer are called monomers.
- A protein is a polymer that consists of individual amino acids linked together.

Carbohydrates, Lipids, and Cholesterol

- Carbohydrates and lipids are organic compounds that contain carbon, hydrogen, and oxygen.
- Lipids store energy in their bonds.
- Unsaturated lipids have one or more double or triple bonds between carbon atoms.
- Cholesterol is a complex lipid.

Self Check

1. **Define** the process by which proteins are made. What other product is formed along with a protein molecule?
2. **Explain** how carbohydrates, proteins, and lipids are important to body functions.
3. **Analyze** how cellulose, starch, and glycogen are different.
4. **Describe** how your body obtains and uses cholesterol.
5. **Think Critically** Explain why even people who eat a healthful diet might gain weight if they don't get enough exercise.

Applying Skills

6. **Draw a Conclusion** Polyunsaturated fats are recommended for a healthful diet. Using what you know about lipids, what might *polyunsaturated* mean?

Looking f🍊r Vitamin C

⬤ Real-World Question

Vitamin C is essential to humans for good health and the prevention of disease. Your body cannot produce this necessary organic molecule, so you must consume it in your food. How do you know which foods are good sources of vitamin C? Reactions that cause color changes are useful as chemical tests. This activity uses the disappearance of the dark-blue color of a solution of starch and iodine to show the presence of vitamin C. Can you test foods for vitamin C? Could the starch-iodine solution be used to show the presence of vitamin C in food?

Goals

- ■ **Prepare** an indicator solution.
- ■ **Verify** a positive test by using a known material.
- ■ **Apply** the test to various foods.
- ■ **Infer** some foods your diet should contain.

Possible Materials

starch solution
iodine solution
vitamin-C solution
water
droppers (10)
15-mL test tubes (10)
test-tube rack
250-mL beaker
stirrer
10-mL graduated cylinder
mortar and pestle
liquid foods such as milk, orange juice, vinegar, and cola
solid foods such as tomatoes, onions, citrus fruits, potatoes, bread, salt, and sugar

Safety Precautions

WARNING: *Do not taste any materials used in the lab. Use care when mashing food samples.*

⬤ Procedure

1. Collect all the materials and equipment you will need.
2. Obtain 10 mL of the starch solution from your teacher.
3. Add the starch solution to 200 mL of water in a 250-mL beaker. Stir.
4. Add four drops of iodine solution to the beaker to make a dark-blue indicator solution. Stir in the drops.

5. Obtain your teacher's approval of your indicator solution before proceeding.

6. **Measure** and place 5 mL of the indicator solution in a clean test tube.

7. Obtain 5 mL of vitamin-C solution from your teacher.

8. Using a clean dropper, add one drop of the vitamin-C solution to the test tube. Stir. Continue adding drops and stirring until you notice a color change. Place a piece of white paper behind the test tube to show the color clearly. Record the number of drops added and any observations.

9. Using a clean test tube and dropper for each test, repeat steps 6 through 8, replacing the vitamin-C solution with other liquids and solids. Add drops of liquid foods or juices until a color change is noted or until you have added about 40 drops of the liquid. Mash solid foods such as onion and potato. Add about 1 g of the food directly to the test tube and stir. Test at least four liquids and four solids.

10. Record the amount of each food added and observations in a table.

◉ *Analyze Your Data*

1. **Infer** What indicates a positive test for vitamin C? How do you know?

2. **Describe** a negative test for vitamin C.

3. **Observe** Which foods tested positive for vitamin C? Which foods, if any, tested negative for vitamin C?

◉ *Conclude and Apply*

1. **Explain** which foods you might include in your diet to make sure you get vitamin C every day.

2. **Determine** if a vitamin-C tablet could take the place of these foods. Explain.

Compare your results with other class members. Were your results consistent? Make a record of the foods you eat for two days. Does your diet contain the minimum RDA of vitamin C?

From Plants to Medicine

Wild plants help save lives

Look carefully at those plants growing in your backyard or neighborhood. With help from scientists, they could save a life. Many of the medicines that doctors prescribe were first developed from plants. For example, aspirin was extracted from the bark of a willow tree. A cancer medication was extracted from the bark of the Pacific yew tree. Aspirin and the cancer medication are now made synthetically—their carbon structures are duplicated in the lab and factory.

Throughout history, and in all parts of the world, traditonal healers have used different parts of plants and flowers to help treat people. Certain kinds of plants have been mashed up and applied to the body to heal burns and sores, or have been swallowed or chewed to help people with illnesses.

Modern researchers are studying the medicinal value of plants and then figuring out the plants' properties and makeup. This is giving scientists important information as they turn to more and more plants to help make medicine in the lab. Studying these plants—and how people in different cultures use them—is the work of scientists called ethnobotanists (eth noh BAH tuhn ihsts). Ethnobotanist Memory Elvin-Lewis notes that plants help treat illnesses.

She visits healers who show her the plants that they find most useful. "Plants are superior chemists producing substances with sophisticated molecular structures that protect the plant from injury and disease," writes Professor Michele L. Trankina. It's these substances in plants that are used as sources of medicines. And it's these substances that are giving researchers and chemists leads to making similar substances in the lab. That can only mean good news—and better health—for people!

Promising cancer medications are made from the bark of the Pacific yew tree.

Memory Elvin-Lewis has spent part of her career studying herbal medicines.

Investigate Research the work of people like Carole L. Cramer. She's modifying common farm plants so that they produce human antibodies used to treat human illnesses. Use the link on the right or your school's media center to get started in your search.

For more information, visit bookl.msscience.com/time

Reviewing Main Ideas

Section 1 Simple Organic Compounds

1. Hydrocarbons are compounds containing only carbon and hydrogen.

2. If a hydrocarbon has only single bonds, it is called a saturated hydrocarbon.

3. Unsaturated hydrocarbons have one or more double or triple bonds in their structure.

Section 2 Other Organic Compounds

1. Hydrogens can be substituted with other atoms or with groups of atoms.

2. An amino acid contains an amino group and a carboxyl group substituted on the same carbon atom.

3. An alcohol is formed when a hydroxyl group is substituted for a hydrogen atom in a hydrocarbon.

4. A carboxylic acid is made when a carboxyl group is substituted and an amine is formed when an amino group ($-NH_2$) is substituted.

Section 3 Biological Compounds

1. Many biological compounds are large molecules called polymers.

2. Proteins serve a variety of functions, including catalyzing many cell reactions.

3. Carbohydrates and lipids are energy sources and the means of storing energy.

Visualizing Main Ideas

Copy and complete the following concept map on simple organic compounds.

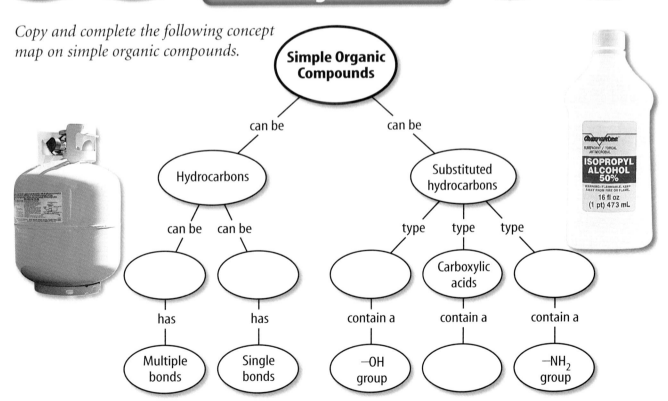

Using Vocabulary

Answer the following questions using complete sentences.

1. Explain the difference between an amino group and an amino acid.

2. How does a hydroxyl group differ from a carboxyl group?

3. Explain why eating carbohydrates would be beneficial to an athlete before a race.

4. What is the connection between a polymer and a protein?

5. What do carbohydrates and lipids have in common?

6. Explain the difference between a saturated and an unsaturated compound.

Checking Concepts

Choose the word or phrase that best answers the question.

7. A certain carbohydrate molecule has ten oxygen atoms. How many hydrogen atoms does it contain?
   A) five
   B) 20
   C) ten
   D) 16

8. Which is NOT a group that can be substituted in a hydrocarbon?
   A) amino
   B) carboxyl
   C) hydroxyl
   D) lipid

9. Which chemical formula represents an alcohol?
   A) CH_3COOH
   B) CH_3NH_2
   C) CH_3OH
   D) CH_4

10. Which substance can build up in arteries and lead to heart disease?
    A) cholesterol
    B) fructose
    C) glucose
    D) starch

11. What is an organic molecule that contains a triple bond called?
    A) polymer
    B) saturated hydrocarbon
    C) isomer
    D) unsaturated hydrocarbon

12. What is the name of the substituted hydrocarbon with the chemical formula CH_2F_2?
    A) methane
    B) fluoromethane
    C) difluoromethane
    D) trifluoromethane

$$F - \underset{\underset{H}{|}}{\overset{\overset{H}{|}}{C}} - F$$

13. Which chemical formula below represents an amino acid?
    A) CH_3COOH
    B) CH_3NH_2
    C) NH_2CH_2COOH
    D) CH_4

14. Proteins are biological polymers made up of what type of monomers?
    A) alcohols
    B) amino acids
    C) ethene molecules
    D) propene molecules

15. Excess energy is stored in your body as which of the following?
    A) proteins
    B) isomers
    C) lipids
    D) saturated hydrocarbons

16. Which is a ring-shaped molecule?
    A) acetone
    B) ethylene
    C) cyclopentane
    D) dichloroethane

Science Online bookl.msscience.com/vocabulary_puzzlemaker

Thinking Critically

Use the figures below to answer question 17.

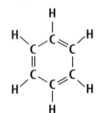

17. **Compare and Contrast** Benzene and cyclohexane are both ring molecules. Discuss the similarities and differences of the two molecules.

18. **Explain** Ethanol is used as a fuel for cars. Explain how energy is obtained from ethanol to fuel a car.

19. **Analyze** Candle wax is one of the longer hydrocarbons. Explain why heat and light are produced in a burning candle.

20. **Infer** In the polymerization of amino acids to make proteins, water molecules are produced as part of the reaction. However, in the polymerization of ethylene, no water is produced. Explain.

21. **Recognize Cause and Effect** Marathon runners go through a process known as hitting the wall. They have used up all their readily available glucose and start using stored lipids as fuel. What is the advantage of eating lots of carbohydrates the day before a race?

22. **Hypothesize** PKU is a genetic disorder that can lead to brain damage. People with this disorder cannot process one of the amino acids. Luckily, damage can be prevented by a proper diet. How is this possible?

23. **Explain** Medicines previously obtained from plants are now manufactured. Can these two medicines be the same?

Performance Activities

24. **Scientific Drawing** Research an amino acid that was not mentioned in the chapter. Draw its structural formula and highlight the portion that substitutes for a hydrogen atom.

Applying Math

Use the graph below to answer question 25.

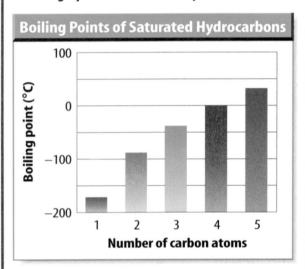

25. **Hydrocarbons** Using the graph above, explain how the boiling point varies with the number of carbon atoms. What do you predict would be the approximate boiling point of hexane?

Use the figure below to answer question 26.

Sucrose

26. **Simple Sugar** What are the percentages of carbon, oxygen, and hydrogen in a sucrose molecule?

27. **Polyethylene** If one polyethylene molecule contains 10,000 ethylene units, how many can be made from 3 million ethylene units?

Part 1 Multiple Choice

Record your answers on the answer sheet provided by your teacher or on a sheet of paper.

Use the table below to answer questions 1 and 2.

Double Bonded Hydrocarbons			
Name	**Formula**	**Name**	**Formula**
Ethene	C_2H_4	Pentene	C_5H_{10}
Propene	C_3H_6	Octene	?
Butene	C_4H_8	Decene	$C_{10}H_{20}$

1. What is the general formula for this family?
   A. $C_{2n}H_n$ **C.** C_nH_{2n}
   B. C_nH_{2n+2} **D.** C_nH_{2n-2}

2. What is the formula of octene?
   A. C_6H_{12} **C.** C_6H_{10}
   B. C_8H_{16} **D.** C_8H_{18}

3. Based on its root name and suffix, what is the structural formula of propyne?
   A. $H-C{\equiv}C-CH_3$
   B. $CH_3-CH_2-CH_3$
   C. $H_2C{=}CHCH_3$
   D. $HC{\equiv}CH$

4. As five amino acids polymerize to form a protein, how many water molecules split off?
   A. 6   **C.** 4
   B. 5   **D.** none

5. As a NH_2 group replaces a hydrogen in a hydrocarbon, which type of compound is formed?
   A. carboxylic acid   **C.** alcohol
   B. amino acid   **D.** amine

6. One of the freons used in refrigerators is dichloro-difluoromethane. How many H atoms are in this molecule ?
   A. 4   **C.** 1
   B. 2   **D.** none

Use the structures below to answer questions 7–9.

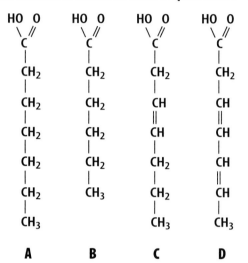

A	**B**	**C**	**D**

7. Which is saturated and has the fewest number of carbon atoms?
   A. A   **C.** C
   B. B   **D.** D

8. Which is a polyunsaturated acid?
   A. A   **C.** C
   B. B   **D.** D

9. These are all considered to be carboxylic acids because they contain which of the following?
   A. a $-COOH$ group
   B. a $-CH_3$ group
   C. a double bond
   D. C, H, and O atoms

Test-Taking Tip

Figures and Illustrations Be sure you understand all symbols in a figure or illustration before attempting to answer any questions about them.

Question 7 Even though the hydrogen molecules are shown on the same side of the carboxylic acid molecules, the structural formula places one hydrogen on either side of the carbon.

Part 2 | Short Response/Grid In

Record answers on the answer sheet provided by your teacher or on a sheet of paper.

Use the illustration below to answer questions 10 and 11.

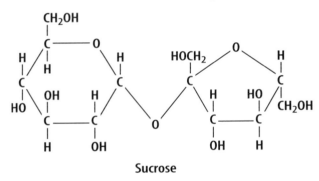

Sucrose

10. If its formula is given as $C_xH_yO_z$ what are the values of *x, y,* and *z*?

11. Sucrose is a carbohydrate. What ratio between atoms denotes a carbohydrate?

12. What is the structural formula of the propyl group?

13. What is the molecular formula of propylamine?

Use the following explanation to answer questions 14 and 15.

A carbon atom has a mass of 12 units and hydrogen 1. A molecule of methane has a molecular mass of $(12 \times 1) + (1 \times 4) = 16$ units, and is $\frac{4}{16} = 25\%$ hydrogen by mass.

Methane

Ethane

14. What is molecular mass of C_2H_6?

15. What is the percent carbon by mass for methane and ethane?

Part 3 | Open Ended

Record your answers on a sheet of paper.

Use the following explanation to answer questions 16–18.

Glycol is a chief component of antifreeze. It IUPAC name is 1, 2-ethanediol.

16. The root "eth" indicates how many carbon atoms?

17. The "ane" suffix indicates which bond between carbons: single, double, or triple?

18. What functional group does "ol" indicate?

Use the table below to answer questions 19 and 20.

Hydrocarbon Isomers	
Formula	**Number of Isomers**
C_2H_6	1
C_4H_{10}	2
C_6H_{14}	5
C_8H_{18}	18
$C_{10}H_{22}$	75

19. Sketch a graph of this data. Is it linear?

20. Predict how many isomers of C_5H_{12} might exist. Draw them.

21. Draw a reasonable structural formula for carbon dioxide. Recall how many bonds each carbon atom can form.

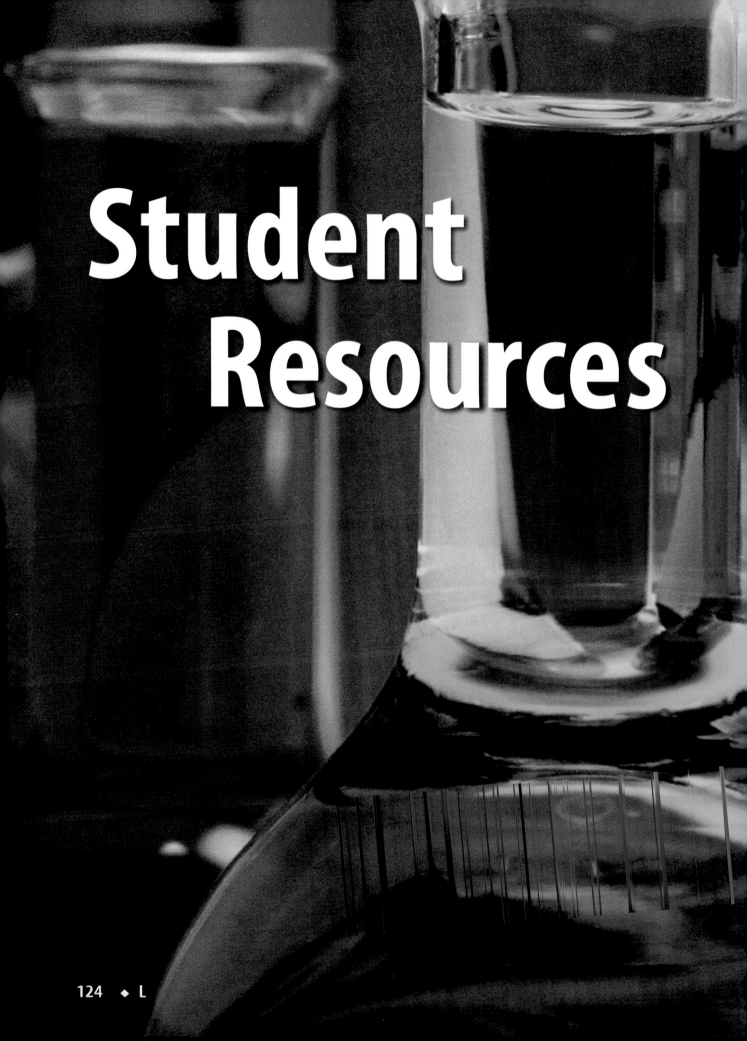

Student Resources

CONTENTS

Scientific Methods

Scientists use an orderly approach called the scientific method to solve problems. This includes organizing and recording data so others can understand them. Scientists use many variations in this method when they solve problems.

Identify a Question

The first step in a scientific investigation or experiment is to identify a question to be answered or a problem to be solved. For example, you might ask which gasoline is the most efficient.

Gather and Organize Information

After you have identified your question, begin gathering and organizing information. There are many ways to gather information, such as researching in a library, interviewing those knowledgeable about the subject, testing and working in the laboratory and field. Fieldwork is investigations and observations done outside of a laboratory.

Researching Information Before moving in a new direction, it is important to gather the information that already is known about the subject. Start by asking yourself questions to determine exactly what you need to know. Then you will look for the information in various reference sources, like the student is doing in **Figure 1.** Some sources may include textbooks, encyclopedias, government documents, professional journals, science magazines, and the Internet. Always list the sources of your information.

Figure 1 The Internet can be a valuable research tool.

Evaluate Sources of Information Not all sources of information are reliable. You should evaluate all of your sources of information, and use only those you know to be dependable. For example, if you are researching ways to make homes more energy efficient, a site written by the U.S. Department of Energy would be more reliable than a site written by a company that is trying to sell a new type of weatherproofing material. Also, remember that research always is changing. Consult the most current resources available to you. For example, a 1985 resource about saving energy would not reflect the most recent findings.

Sometimes scientists use data that they did not collect themselves, or conclusions drawn by other researchers. This data must be evaluated carefully. Ask questions about how the data were obtained, if the investigation was carried out properly, and if it has been duplicated exactly with the same results. Would you reach the same conclusion from the data? Only when you have confidence in the data can you believe it is true and feel comfortable using it.

Interpret Scientific Illustrations As you research a topic in science, you will see drawings, diagrams, and photographs to help you understand what you read. Some illustrations are included to help you understand an idea that you can't see easily by yourself, like the tiny particles in an atom in **Figure 2.** A drawing helps many people to remember details more easily and provides examples that clarify difficult concepts or give additional information about the topic you are studying. Most illustrations have labels or a caption to identify or to provide more information.

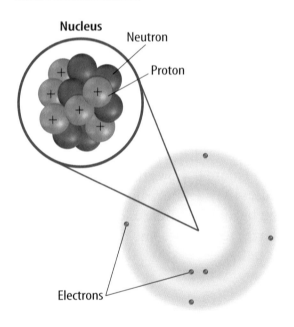

Figure 2 This drawing shows an atom of carbon with its six protons, six neutrons, and six electrons.

Concept Maps One way to organize data is to draw a diagram that shows relationships among ideas (or concepts). A concept map can help make the meanings of ideas and terms more clear, and help you understand and remember what you are studying. Concept maps are useful for breaking large concepts down into smaller parts, making learning easier.

Network Tree A type of concept map that not only shows a relationship, but how the concepts are related is a network tree, shown in **Figure 3.** In a network tree, the words are written in the ovals, while the description of the type of relationship is written across the connecting lines.

When constructing a network tree, write down the topic and all major topics on separate pieces of paper or notecards. Then arrange them in order from general to specific. Branch the related concepts from the major concept and describe the relationship on the connecting line. Continue to more specific concepts until finished.

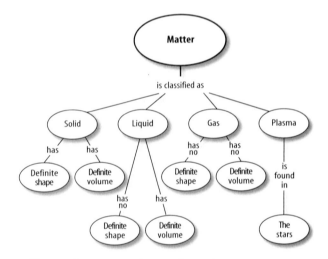

Figure 3 A network tree shows how concepts or objects are related.

Events Chain Another type of concept map is an events chain. Sometimes called a flow chart, it models the order or sequence of items. An events chain can be used to describe a sequence of events, the steps in a procedure, or the stages of a process.

When making an events chain, first find the one event that starts the chain. This event is called the initiating event. Then, find the next event and continue until the outcome is reached, as shown in **Figure 4.**

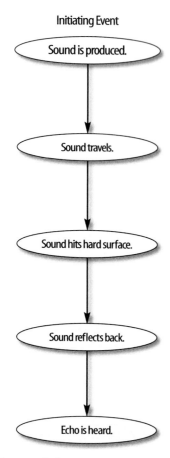

Initiating Event

Sound is produced.

Sound travels.

Sound hits hard surface.

Sound reflects back.

Echo is heard.

Figure 4 Events-chain concept maps show the order of steps in a process or event. This concept map shows how a sound makes an echo.

Cycle Map A specific type of events chain is a cycle map. It is used when the series of events do not produce a final outcome, but instead relate back to the beginning event, such as in **Figure 5.** Therefore, the cycle repeats itself.

To make a cycle map, first decide what event is the beginning event. This is also called the initiating event. Then list the next events in the order that they occur, with the last event relating back to the initiating event. Words can be written between the events that describe what happens from one event to the next. The number of events in a cycle map can vary, but usually contain three or more events.

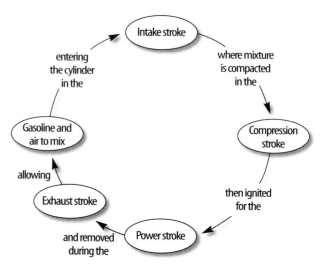

Figure 5 A cycle map shows events that occur in a cycle.

Spider Map A type of concept map that you can use for brainstorming is the spider map. When you have a central idea, you might find that you have a jumble of ideas that relate to it but are not necessarily clearly related to each other. The spider map on sound in **Figure 6** shows that if you write these ideas outside the main concept, then you can begin to separate and group unrelated terms so they become more useful.

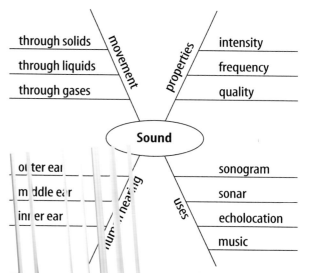

Figure 6 A spider map allows you to list ideas that relate to a central topic but not necessarily to one another.

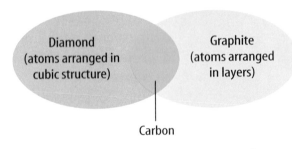

Figure 7 This Venn diagram compares and contrasts two substances made from carbon.

Venn Diagram To illustrate how two subjects compare and contrast you can use a Venn diagram. You can see the characteristics that the subjects have in common and those that they do not, shown in **Figure 7.**

To create a Venn diagram, draw two overlapping ovals that that are big enough to write in. List the characteristics unique to one subject in one oval, and the characteristics of the other subject in the other oval. The characteristics in common are listed in the overlapping section.

Make and Use Tables One way to organize information so it is easier to understand is to use a table. Tables can contain numbers, words, or both.

To make a table, list the items to be compared in the first column and the characteristics to be compared in the first row. The title should clearly indicate the content of the table, and the column or row heads should be clear. Notice that in **Table 1** the units are included.

Table 1 Recyclables Collected During Week			
Day of Week	Paper (kg)	Aluminum (kg)	Glass (kg)
Monday	5.0	4.0	12.0
Wednesday	4.0	1.0	10.0
Friday	2.5	2.0	10.0

Make a Model One way to help you better understand the parts of a structure, the way a process works, or to show things too large or small for viewing is to make a model. For example, an atomic model made of a plastic-ball nucleus and pipe-cleaner electron shells can help you visualize how the parts of an atom relate to each other. Other types of models can by devised on a computer or represented by equations.

Form a Hypothesis

A possible explanation based on previous knowledge and observations is called a hypothesis. After researching gasoline types and recalling previous experiences in your family's car you form a hypothesis—our car runs more efficiently because we use premium gasoline. To be valid, a hypothesis has to be something you can test by using an investigation.

Predict When you apply a hypothesis to a specific situation, you predict something about that situation. A prediction makes a statement in advance, based on prior observation, experience, or scientific reasoning. People use predictions to make everyday decisions. Scientists test predictions by performing investigations. Based on previous observations and experiences, you might form a prediction that cars are more efficient with premium gasoline. The prediction can be tested in an investigation.

Design an Experiment A scientist needs to make many decisions before beginning an investigation. Some of these include: how to carry out the investigation, what steps to follow, how to record the data, and how the investigation will answer the question. It also is important to address any safety concerns.

Test the Hypothesis

Now that you have formed your hypothesis, you need to test it. Using an investigation, you will make observations and collect data, or information. This data might either support or not support your hypothesis. Scientists collect and organize data as numbers and descriptions.

Follow a Procedure In order to know what materials to use, as well as how and in what order to use them, you must follow a procedure. **Figure 8** shows a procedure you might follow to test your hypothesis.

Procedure

1. Use regular gasoline for two weeks.
2. Record the number of kilometers between fill-ups and the amount of gasoline used.
3. Switch to premium gasoline for two weeks.
4. Record the number of kilometers between fill-ups and the amount of gasoline used.

Figure 8 A procedure tells you what to do step by step.

Identify and Manipulate Variables and Controls In any experiment, it is important to keep everything the same except for the item you are testing. The one factor you change is called the independent variable. The change that results is the dependent variable. Make sure you have only one independent variable, to assure yourself of the cause of the changes you observe in the dependent variable. For example, in your gasoline experiment the type of fuel is the independent variable. The dependent variable is the efficiency.

Many experiments also have a control—an individual instance or experimental subject for which the independent variable is not changed. You can then compare the test results to the control results. To design a control you can have two cars of the same type. The control car uses regular gasoline for four weeks. After you are done with the test, you can compare the experimental results to the control results.

Collect Data

Whether you are carrying out an investigation or a short observational experiment, you will collect data, as shown in **Figure 9.** Scientists collect data as numbers and descriptions and organize it in specific ways.

Observe Scientists observe items and events, then record what they see. When they use only words to describe an observation, it is called qualitative data. Scientists' observations also can describe how much there is of something. These observations use numbers, as well as words, in the description and are called quantitative data. For example, if a sample of the element gold is described as being "shiny and very dense" the data are qualitative. Quantitative data on this sample of gold might include "a mass of 30 g and a density of 19.3 g/cm^3."

Figure 9 Collecting data is one way to gather information directly.

Figure 10 Record data neatly and clearly so it is easy to understand.

When you make observations you should examine the entire object or situation first, and then look carefully for details. It is important to record observations accurately and completely. Always record your notes immediately as you make them, so you do not miss details or make a mistake when recording results from memory. Never put unidentified observations on scraps of paper. Instead they should be recorded in a notebook, like the one in **Figure 10.** Write your data neatly so you can easily read it later. At each point in the experiment, record your observations and label them. That way, you will not have to determine what the figures mean when you look at your notes later. Set up any tables that you will need to use ahead of time, so you can record any observations right away. Remember to avoid bias when collecting data by not including personal thoughts when you record observations. Record only what you observe.

Estimate Scientific work also involves estimating. To estimate is to make a judgment about the size or the number of something without measuring or counting. This is important when the number or size of an object or population is too large or too difficult to accurately count or measure.

Sample Scientists may use a sample or a portion of the total number as a type of estimation. To sample is to take a small, representative portion of the objects or organisms of a population for research. By making careful observations or manipulating variables within that portion of the group, information is discovered and conclusions are drawn that might apply to the whole population. A poorly chosen sample can be unrepresentative of the whole. If you were trying to determine the rainfall in an area, it would not be best to take a rainfall sample from under a tree.

Measure You use measurements everyday. Scientists also take measurements when collecting data. When taking measurements, it is important to know how to use measuring tools properly. Accuracy also is important.

Length To measure length, the distance between two points, scientists use meters. Smaller measurements might be measured in centimeters or millimeters.

Length is measured using a metric ruler or meter stick. When using a metric ruler, line up the 0-cm mark with the end of the object being measured and read the number of the unit where the object ends. Look at the metric ruler shown in **Figure 11.** The centimeter lines are the long, numbered lines, and the shorter lines are millimeter lines. In this instance, the length would be 4.50 cm.

Figure 11 This metric ruler has centimeter and millimeter divisions.

Mass The SI unit for mass is the kilogram (kg). Scientists can measure mass using units formed by adding metric prefixes to the unit gram (g), such as milligram (mg). To measure mass, you might use a triple-beam balance similar to the one shown in **Figure 12.** The balance has a pan on one side and a set of beams on the other side. Each beam has a rider that slides on the beam.

When using a triple-beam balance, place an object on the pan. Slide the largest rider along its beam until the pointer drops below zero. Then move it back one notch. Repeat the process for each rider proceeding from the larger to smaller until the pointer swings an equal distance above and below the zero point. Sum the masses on each beam to find the mass of the object. Move all riders back to zero when finished.

Instead of putting materials directly on the balance, scientists often take a tare of a container. A tare is the mass of a container into which objects or substances are placed for measuring their masses. To mass objects or substances, find the mass of a clean container. Remove the container from the pan, and place the object or substances in the container. Find the mass of the container with the materials in it. Subtract the mass of the empty container from the mass of the filled container to find the mass of the materials you are using.

Figure 12 A triple-beam balance is used to determine the mass of an object.

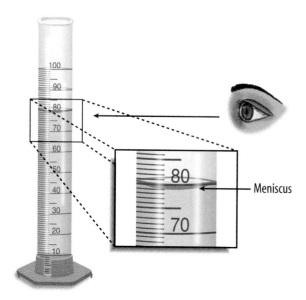

Figure 13 Graduated cylinders measure liquid volume.

Liquid Volume To measure liquids, the unit used is the liter. When a smaller unit is needed, scientists might use a milliliter. Because a milliliter takes up the volume of a cube measuring 1 cm on each side it also can be called a cubic centimeter ($cm^3 = cm \times cm \times cm$).

You can use beakers and graduated cylinders to measure liquid volume. A graduated cylinder, shown in **Figure 13,** is marked from bottom to top in milliliters. In lab, you might use a 10-mL graduated cylinder or a 100-mL graduated cylinder. When measuring liquids, notice that the liquid has a curved surface. Look at the surface at eye level, and measure the bottom of the curve. This is called the meniscus. The graduated cylinder in **Figure 13** contains 79.0 mL, or 79.0 cm^3, of a liquid.

Temperature Scientists often measure temperature using the Celsius scale. Pure water has a freezing point of 0°C and boiling point of 100°C. The unit of measurement is degrees Celsius. Two other scales often used are the Fahrenheit and Kelvin scales.

Figure 14 A thermometer measures the temperature of an object.

Scientists use a thermometer to measure temperature. Most thermometers in a laboratory are glass tubes with a bulb at the bottom end containing a liquid such as colored alcohol. The liquid rises or falls with a change in temperature. To read a glass thermometer like the thermometer in **Figure 14,** rotate it slowly until a red line appears. Read the temperature where the red line ends.

Form Operational Definitions An operational definition defines an object by how it functions, works, or behaves. For example, when you are playing hide and seek and a tree is home base, you have created an operational definition for a tree.

Objects can have more than one operational definition. For example, a ruler can be defined as a tool that measures the length of an object (how it is used). It can also be a tool with a series of marks used as a standard when measuring (how it works).

Analyze the Data

To determine the meaning of your observations and investigation results, you will need to look for patterns in the data. Then you must think critically to determine what the data mean. Scientists use several approaches when they analyze the data they have collected and recorded. Each approach is useful for identifying specific patterns.

Interpret Data The word *interpret* means "to explain the meaning of something." When analyzing data from an experiment, try to find out what the data show. Identify the control group and the test group to see whether or not changes in the independent variable have had an effect. Look for differences in the dependent variable between the control and test groups.

Classify Sorting objects or events into groups based on common features is called classifying. When classifying, first observe the objects or events to be classified. Then select one feature that is shared by some members in the group, but not by all. Place those members that share that feature in a subgroup. You can classify members into smaller and smaller subgroups based on characteristics. Remember that when you classify, you are grouping objects or events for a purpose. Keep your purpose in mind as you select the features to form groups and subgroups.

Compare and Contrast Observations can be analyzed by noting the similarities and differences between two more objects or events that you observe. When you look at objects or events to see how they are similar, you are comparing them. Contrasting is looking for differences in objects or events.

Recognize Cause and Effect A cause is a reason for an action or condition. The effect is that action or condition. When two events happen together, it is not necessarily true that one event caused the other. Scientists must design a controlled investigation to recognize the exact cause and effect.

Draw Conclusions

When scientists have analyzed the data they collected, they proceed to draw conclusions about the data. These conclusions are sometimes stated in words similar to the hypothesis that you formed earlier. They may confirm a hypothesis, or lead you to a new hypothesis.

Infer Scientists often make inferences based on their observations. An inference is an attempt to explain observations or to indicate a cause. An inference is not a fact, but a logical conclusion that needs further investigation. For example, you may infer that a fire has caused smoke. Until you investigate, however, you do not know for sure.

Apply When you draw a conclusion, you must apply those conclusions to determine whether the data supports the hypothesis. If your data do not support your hypothesis, it does not mean that the hypothesis is wrong. It means only that the result of the investigation did not support the hypothesis. Maybe the experiment needs to be redesigned, or some of the initial observations on which the hypothesis was based were incomplete or biased. Perhaps more observation or research is needed to refine your hypothesis. A successful investigation does not always come out the way you originally predicted.

Avoid Bias Sometimes a scientific investigation involves making judgments. When you make a judgment, you form an opinion. It is important to be honest and not to allow any expectations of results to bias your judgments. This is important throughout the entire investigation, from researching to collecting data to drawing conclusions.

Communicate

The communication of ideas is an important part of the work of scientists. A discovery that is not reported will not advance the scientific community's understanding or knowledge. Communication among scientists also is important as a way of improving their investigations.

Scientists communicate in many ways, from writing articles in journals and magazines that explain their investigations and experiments, to announcing important discoveries on television and radio. Scientists also share ideas with colleagues on the Internet or present them as lectures, like the student is doing in **Figure 15.**

Figure 15 A student communicates to his peers about his investigation.

SAFETY SYMBOLS

SAFETY SYMBOLS	HAZARD	EXAMPLES	PRECAUTION	REMEDY
DISPOSAL	Special disposal procedures need to be followed.	certain chemicals, living organisms	Do not dispose of these materials in the sink or trash can.	Dispose of wastes as directed by your teacher.
BIOLOGICAL	Organisms or other biological materials that might be harmful to humans	bacteria, fungi, blood, unpreserved tissues, plant materials	Avoid skin contact with these materials. Wear mask or gloves.	Notify your teacher if you suspect contact with material. Wash hands thoroughly.
EXTREME TEMPERATURE	Objects that can burn skin by being too cold or too hot	boiling liquids, hot plates, dry ice, liquid nitrogen	Use proper protection when handling.	Go to your teacher for first aid.
SHARP OBJECT	Use of tools or glassware that can easily puncture or slice skin	razor blades, pins, scalpels, pointed tools, dissecting probes, broken glass	Practice common-sense behavior and follow guidelines for use of the tool.	Go to your teacher for first aid.
FUME	Possible danger to respiratory tract from fumes	ammonia, acetone, nail polish remover, heated sulfur, moth balls	Make sure there is good ventilation. Never smell fumes directly. Wear a mask.	Leave foul area and notify your teacher immediately.
ELECTRICAL	Possible danger from electrical shock or burn	improper grounding, liquid spills, short circuits, exposed wires	Double-check setup with teacher. Check condition of wires and apparatus.	Do not attempt to fix electrical problems. Notify your teacher immediately.
IRRITANT	Substances that can irritate the skin or mucous membranes of the respiratory tract	pollen, moth balls, steel wool, fiberglass, potassium permanganate	Wear dust mask and gloves. Practice extra care when handling these materials.	Go to your teacher for first aid.
CHEMICAL	Chemicals can react with and destroy tissue and other materials	bleaches such as hydrogen peroxide; acids such as sulfuric acid, hydrochloric acid; bases such as ammonia, sodium hydroxide	Wear goggles, gloves, and an apron.	Immediately flush the affected area with water and notify your teacher.
TOXIC	Substance may be poisonous if touched, inhaled, or swallowed.	mercury, many metal compounds, iodine, poinsettia plant parts	Follow your teacher's instructions.	Always wash hands thoroughly after use. Go to your teacher for first aid.
FLAMMABLE	Flammable chemicals may be ignited by open flame, spark, or exposed heat.	alcohol, kerosene, potassium permanganate	Avoid open flames and heat when using flammable chemicals.	Notify your teacher immediately. Use fire safety equipment if applicable.
OPEN FLAME	Open flame in use, may cause fire.	hair, clothing, paper, synthetic materials	Tie back hair and loose clothing. Follow teacher's instruction on lighting and extinguishing flames.	Notify your teacher immediately. Use fire safety equipment if applicable.

 Eye Safety Proper eye protection should be worn at all times by anyone performing or observing science activities.

 Clothing Protection This symbol appears when substances could stain or burn clothing.

 Animal Safety This symbol appears when safety of animals and students must be ensured.

 Handwashing After the lab, wash hands with soap and water before removing goggles.

Safety in the Science Laboratory

The science laboratory is a safe place to work if you follow standard safety procedures. Being responsible for your own safety helps to make the entire laboratory a safer place for everyone. When performing any lab, read and apply the caution statements and safety symbol listed at the beginning of the lab.

General Safety Rules

1. Obtain your teacher's permission to begin all investigations and use laboratory equipment.

2. Study the procedure. Ask your teacher any questions. Be sure you understand safety symbols shown on the page.

3. Notify your teacher about allergies or other health conditions which can affect your participation in a lab.

4. Learn and follow use and safety procedures for your equipment. If unsure, ask your teacher.

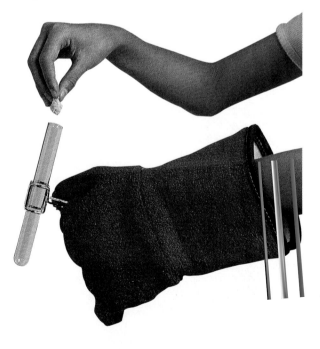

5. Never eat, drink, chew gum, apply cosmetics, or do any personal grooming in the lab. Never use lab glassware as food or drink containers. Keep your hands away from your face and mouth.

6. Know the location and proper use of the safety shower, eye wash, fire blanket, and fire alarm.

Prevent Accidents

1. Use the safety equipment provided to you. Goggles and a safety apron should be worn during investigations.

2. Do NOT use hair spray, mousse, or other flammable hair products. Tie back long hair and tie down loose clothing.

3. Do NOT wear sandals or other open-toed shoes in the lab.

4. Remove jewelry on hands and wrists. Loose jewelry, such as chains and long necklaces, should be removed to prevent them from getting caught in equipment.

5. Do not taste any substances or draw any material into a tube with your mouth.

6. Proper behavior is expected in the lab. Practical jokes and fooling around can lead to accidents and injury.

7. Keep your work area uncluttered.

Laboratory Work

1. Collect and carry all equipment and materials to your work area before beginning a lab.

2. Remain in your own work area unless given permission by your teacher to leave it.

3. Dispose of chemicals and other materials as directed by your teacher. Place broken glass and solid substances in the proper containers. Never discard materials in the sink.

4. Clean your work area.

5. Wash your hands with soap and water thoroughly BEFORE removing your goggles.

Emergencies

1. Report any fire, electrical shock, glassware breakage, spill, or injury, no matter how small, to your teacher immediately. Follow his or her instructions.

2. If your clothing should catch fire, STOP, DROP, and ROLL. If possible, smother it with the fire blanket or get under a safety shower. NEVER RUN.

3. If a fire should occur, turn off all gas and leave the room according to established procedures.

4. In most instances, your teacher will clean up spills. Do NOT attempt to clean up spills unless you are given permission and instructions to do so.

5. If chemicals come into contact with your eyes or skin, notify your teacher immediately. Use the eyewash or flush your skin or eyes with large quantities of water.

6. The fire extinguisher and first-aid kit should only be used by your teacher unless it is an extreme emergency and you have been given permission.

7. If someone is injured or becomes ill, only a professional medical provider or someone certified in first aid should perform first-aid procedures.

3. Always slant test tubes away from yourself and others when heating them, adding substances to them, or rinsing them.

4. If instructed to smell a substance in a container, hold the container a short distance away and fan vapors towards your nose.

5. Do NOT substitute other chemicals/substances for those in the materials list unless instructed to do so by your teacher.

6. Do NOT take any materials or chemicals outside of the laboratory.

7. Stay out of storage areas unless instructed to be there and supervised by your teacher.

Laboratory Cleanup

1. Turn off all burners, water, and gas, and disconnect all electrical devices.

2. Clean all pieces of equipment and return all materials to their proper places.

EXTRA  Labs

From Your Kitchen, Junk Drawer, or Yard

1 Human Bonding

▶ Real-World Question
How can humans model atoms bonding together?

Possible Materials 🔲
- family members or friends
- sheets of blank paper
- markers
- large safety pins
- large colored rubber bands

▶ Procedure
1. Draw a large electron dot diagram of an element you choose. Have other activity participants do that too.
2. Pin the diagram to your shirt.
3. How many electrons does your element have? Gather that many rubber bands.
4. Place about half of the rubber bands on one wrist and half on the other.
5. Form bonds by finding someone who has the number of rubber bands you need to total eight. Try to form as many different compounds with different elements as you can. (You may need two or three of another element's atoms to make a compound.) Record the compounds you make in your Science Journal. Label each compound as ionic or covalent.

▶ Conclude and Apply
1. Which elements don't form any bonds?
2. Which elements form four bonds?

2 Mini Fireworks

▶ Real-World Question
Where do the colors in fireworks come from?

Possible Materials 🔲 🔲 🔲 🔲 🔲
- candle
- lighter
- wooden chopsticks (or a fork or tongs)
- penny
- water in an old cup
- steel wool

▶ Procedure
1. Light the candle.
2. Use the chopsticks to get a firm grip on the penny.
3. Hold the penny in the flame until you observe a change. (Hint: this experiment is more fun in the bathroom with the lights off!)
4. Drop the penny in the water when you are finished and plunge the burning end of the chopsticks or hot part of the fork into the water as well.
5. Repeat the procedure using steel wool.

▶ Conclude and Apply
1. What color do you see?
2. Infer why copper and iron are used in fireworks.
3. Research what other elements are used in fireworks.

Adult supervision required for all labs.

3 A Good Mix?

▶ **Real-World Question**

What liquids will dissolve in water?

Possible Materials
- cooking oil
- water
- apple or grape juice
- rubbing alcohol
- spoon
- glass
- measuring cup

▶ **Procedure**

1. Pour 100 mL of water into a large glass.
2. Pour 100 mL of apple juice into the glass and stir the water and juice together. Observe your mixture to determine whether juice is soluble in water.

3. Empty and rinse out your glass.
4. Pour 100 mL of water and 100 mL of cooking oil into the glass and stir them together. Observe your mixture to determine whether oil is soluble in water.
5. Empty and rinse out your glass.
6. Pour 100 mL of water and 100 mL of rubbing alcohol into the glass and stir them together. Observe your mixture to determine whether alcohol is soluble in water.

▶ **Conclude and Apply**

1. List the liquid(s) that are soluble in water.
2. List the liquid(s) that are not soluble in water.
3. Infer why some liquids are soluble in water and others are not.

4 Liquid Lab

▶ **Real-World Question**

How do the properties of water and rubbing alcohol compare?

Possible Materials
- water
- rubbing alcohol
- vegetable oil
- glasses (2)
- ice cubes (2)
- measuring cup
- spoon

▶ **Procedure**

1. Copy the Physical Properties chart into your Science Journal.
2. Slowly pour 200 mL of water into one glass and observe the viscosity of water. Slowly pour 200 mL of rubbing alcohol into a second glass and observe the viscosity of rubbing alcohol. Record your observations in your chart.

3. Observe the color and odor of both liquids and record your observations in your chart.
4. Drop an ice cube into each glass. Comment on the density of each liquid in your chart.
5. Pour 50 mL of vegetable oil into each glass and stir. Record your observations in your chart.

▶ **Conclude and Apply**

1. Compare the properties of water and isopropyl alcohol.
2. Infer what would happen if water and isopropyl alcohol were mixed together.

Physical Properties		
	Water	Rubbing Alcohol
Color		
Odor		
Viscosity		
Density		
Solubility with oil		

Computer Skills

People who study science rely on computers, like the one in **Figure 16,** to record and store data and to analyze results from investigations. Whether you work in a laboratory or just need to write a lab report with tables, good computer skills are a necessity.

Using the computer comes with responsibility. Issues of ownership, security, and privacy can arise. Remember, if you did not author the information you are using, you must provide a source for your information. Also, anything on a computer can be accessed by others. Do not put anything on the computer that you would not want everyone to know. To add more security to your work, use a password.

Use a Word Processing Program

A computer program that allows you to type your information, change it as many times as you need to, and then print it out is called a word processing program. Word processing programs also can be used to make tables.

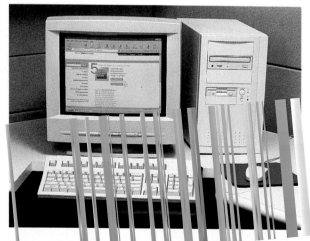

Figure 16 A computer will make reports neater and more professional looking.

Learn the Skill To start your word processing program, a blank document, sometimes called "Document 1," appears on the screen. To begin, start typing. To create a new document, click the *New* button on the standard tool bar. These tips will help you format the document.

- The program will automatically move to the next line; press *Enter* if you wish to start a new paragraph.
- Symbols, called non-printing characters, can be hidden by clicking the *Show/Hide* button on your toolbar.
- To insert text, move the cursor to the point where you want the insertion to go, click on the mouse once, and type the text.
- To move several lines of text, select the text and click the *Cut* button on your toolbar. Then position your cursor in the location that you want to move the cut text and click *Paste.* If you move to the wrong place, click *Undo.*
- The spell check feature does not catch words that are misspelled to look like other words, like "cold" instead of "gold." Always reread your document to catch all spelling mistakes.
- To learn about other word processing methods, read the user's manual or click on the *Help* button.
- You can integrate databases, graphics, and spreadsheets into documents by copying from another program and pasting it into your document, or by using desktop publishing (DTP). DTP software allows you to put text and graphics together to finish your document with a professional look. This software varies in how it is used and its capabilities.

Use a Database

A collection of facts stored in a computer and sorted into different fields is called a database. A database can be reorganized in any way that suits your needs.

Learn the Skill A computer program that allows you to create your own database is a database management system (DBMS). It allows you to add, delete, or change information. Take time to get to know the features of your database software.

- Determine what facts you would like to include and research to collect your information.
- Determine how you want to organize the information.
- Follow the instructions for your particular DBMS to set up fields. Then enter each item of data in the appropriate field.
- Follow the instructions to sort the information in order of importance.
- Evaluate the information in your database, and add, delete, or change as necessary.

Use the Internet

The Internet is a global network of computers where information is stored and shared. To use the Internet, like the students in **Figure 17,** you need a modem to connect your computer to a phone line and an Internet Service Provider account.

Learn the Skill To access internet sites and information, use a "Web browser," which lets you view and explore pages on the World Wide Web. Each page is its own site, and each site has its own address, called a URL. Once you have found a Web browser, follow these steps for a search (this also is how you search a database).

Figure 17 The Internet allows you to search a global network for a variety of information.

- Be as specific as possible. If you know you want to research "gold," don't type in "elements." Keep narrowing your search until you find what you want.
- Web sites that end in *.com* are commercial Web sites; *.org, .edu,* and *.gov* are non-profit, educational, or government Web sites.
- Electronic encyclopedias, almanacs, indexes, and catalogs will help locate and select relevant information.
- Develop a "home page" with relative ease. When developing a Web site, NEVER post pictures or disclose personal information such as location, names, or phone numbers. Your school or community usually can host your Web site. A basic understanding of HTML (hypertext mark-up language), the language of Web sites, is necessary. Software that creates HTML code is called authoring software, and can be downloaded free from many Web sites. This software allows text and pictures to be arranged as the software is writing the HTML code.

Use a Spreadsheet

A spreadsheet, shown in **Figure 18,** can perform mathematical functions with any data arranged in columns and rows. By entering a simple equation into a cell, the program can perform operations in specific cells, rows, or columns.

Learn the Skill Each column (vertical) is assigned a letter, and each row (horizontal) is assigned a number. Each point where a row and column intersect is called a cell, and is labeled according to where it is located—Column A, Row 1 (A1).

- Decide how to organize the data, and enter it in the correct row or column.
- Spreadsheets can use standard formulas or formulas can be customized to calculate cells.
- To make a change, click on a cell to make it activate, and enter the edited data or formula.
- Spreadsheets also can display your results in graphs. Choose the style of graph that best represents the data.

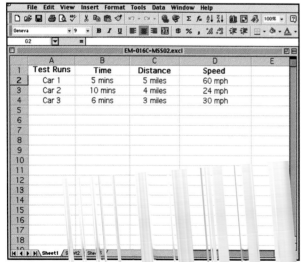

Figure 18 A spreadsheet allows you perform mathematical operations on your data.

Use Graphics Software

Adding pictures, called graphics, to your documents is one way to make your documents more meaningful and exciting. This software adds, edits, and even constructs graphics. There is a variety of graphics software programs. The tools used for drawing can be a mouse, keyboard, or other specialized devices. Some graphics programs are simple. Others are complicated, called computer-aided design (CAD) software.

Learn the Skill It is important to have an understanding of the graphics software being used before starting. The better the software is understood, the better the results. The graphics can be placed in a word-processing document.

- Clip art can be found on a variety of internet sites, and on CDs. These images can be copied and pasted into your document.
- When beginning, try editing existing drawings, then work up to creating drawings.
- The images are made of tiny rectangles of color called pixels. Each pixel can be altered.
- Digital photography is another way to add images. The photographs in the memory of a digital camera can be downloaded into a computer, then edited and added to the document.
- Graphics software also can allow a in a tin. The software allow drawings to h e the appeara o move ner y connecting bsic w s auom ca ly Ths is called in-l ve ng o t e ni g Remember to av te

Presentation Skills

Develop Multimedia Presentations

Most presentations are more dynamic if they include diagrams, photographs, videos, or sound recordings, like the one shown in **Figure 19.** A multimedia presentation involves using stereos, overhead projectors, televisions, computers, and more.

Learn the Skill Decide the main points of your presentation, and what types of media would best illustrate those points.

- Make sure you know how to use the equipment you are working with.
- Practice the presentation using the equipment several times.
- Enlist the help of a classmate to push play or turn lights out for you. Be sure to practice your presentation with him or her.
- If possible, set up all of the equipment ahead of time, and make sure everything is working properly.

Figure 19 These students are engaging the audience using a variety of tools.

Computer Presentations

There are many different interactive computer programs that you can use to enhance your presentation. Most computers have a compact disc (CD) drive that can play both CDs and digital video discs (DVDs). Also, there is hardware to connect a regular CD, DVD, or VCR. These tools will enhance your presentation.

Another method of using the computer to aid in your presentation is to develop a slide show using a computer program. This can allow movement of visuals at the presenter's pace, and can allow for visuals to build on one another.

Learn the Skill In order to create multimedia presentations on a computer, you need to have certain tools. These may include traditional graphic tools and drawing programs, animation programs, and authoring systems that tie everything together. Your computer will tell you which tools it supports. The most important step is to learn about the tools that you will be using.

- Often, color and strong images will convey a point better than words alone. Use the best methods available to convey your point.
- As with other presentations, practice many times.
- Practice your presentation with the tools you and any assistants will be using.
- Maintain eye contact with the audience. The purpose of using the computer is not to prompt the presenter, but to help the audience understand the points of the presentation.

Math Review

Use Fractions

A fraction compares a part to a whole. In the fraction $\frac{2}{3}$, the 2 represents the part and is the numerator. The 3 represents the whole and is the denominator.

Reduce Fractions To reduce a fraction, you must find the largest factor that is common to both the numerator and the denominator, the greatest common factor (GCF). Divide both numbers by the GCF. The fraction has then been reduced, or it is in its simplest form.

Example Twelve of the 20 chemicals in the science lab are in powder form. What fraction of the chemicals used in the lab are in powder form?

Step 1 Write the fraction.

$$\frac{part}{whole} = \frac{12}{20}$$

Step 2 To find the GCF of the numerator and denominator, list all of the factors of each number.

Factors of 12: 1, 2, 3, 4, 6, 12 (the numbers that divide evenly into 12)

Factors of 20: 1, 2, 4, 5, 10, 20 (the numbers that divide evenly into 20)

Step 3 List the common factors.

1, 2, 4.

Step 4 Choose the greatest factor in the list.

The GCF of 12 and 20 is 4.

Step 5 Divide the numerator and denominator by the GCF.

$$\frac{12 \div 4}{20 \div 4} = \frac{3}{5}$$

In the lab, $\frac{3}{5}$ of the chemicals are in powder form

Practice Problem At an amusement park, 66 of 90 rides have a height restriction. What fraction of the rides, in its simplest form, has a height restriction?

Add and Subtract Fractions To add or subtract fractions with the same denominator, add or subtract the numerators and write the sum or difference over the denominator. After finding the sum or difference, find the simplest form for your fraction.

Example 1 In the forest outside your house, $\frac{1}{8}$ of the animals are rabbits, $\frac{3}{8}$ are squirrels, and the remainder are birds and insects. How many are mammals?

Step 1 Add the numerators.

$$\frac{1}{8} + \frac{3}{8} = \frac{(1 + 3)}{8} = \frac{4}{8}$$

Step 2 Find the GCF.

$$\frac{4}{8} \quad (\text{GCF, 4})$$

Step 3 Divide the numerator and denominator by the GCF.

$$\frac{4}{4} = 1, \ \frac{8}{4} = 2$$

$\frac{1}{2}$ of the animals are mammals.

Example 2 If $\frac{7}{16}$ of the Earth is covered by freshwater, and $\frac{1}{16}$ of that is in glaciers, how much freshwater is not frozen?

Step 1 Subtract the numerators.

$$\frac{7}{16} - \frac{1}{16} = \frac{(7 - 1)}{16} = \frac{6}{16}$$

Step 2 Find the GCF.

$$\frac{6}{16} \quad (\text{GCF, 2})$$

Step 3 Divide the numerator and denominator by the GCF.

$$\frac{6}{2} = 3, \ \frac{16}{2} = 8$$

$\frac{3}{8}$ of the fresh water is not frozen.

Practice Problem A bicycle rider is going 15 km/h for $\frac{4}{9}$ of his ride, 10 km/h for $\frac{2}{9}$ of his ride, and 8 km/h for the remainder of the ride. How much of his ride is he going over 8 km/h?

Unlike Denominators To add or subtract fractions with unlike denominators, first find the least common denominator (LCD). This is the smallest number that is a common multiple of both denominators. Rename each fraction with the LCD, and then add or subtract. Find the simplest form if necessary.

Example 1 A chemist makes a paste that is $\frac{1}{2}$ table salt (NaCl), $\frac{1}{3}$ sugar ($C_6H_{12}O_6$), and the rest water (H_2O). How much of the paste is a solid?

Step 1 Find the LCD of the fractions.

$\frac{1}{2} + \frac{1}{3}$ (LCD, 6)

Step 2 Rename each numerator and each denominator with the LCD.

$1 \times 3 = 3, \quad 2 \times 3 = 6$

$1 \times 2 = 2, \quad 3 \times 2 = 6$

Step 3 Add the numerators.

$\frac{3}{6} + \frac{2}{6} = \frac{(3 + 2)}{6} = \frac{5}{6}$

$\frac{5}{6}$ of the paste is a solid.

Example 2 The average precipitation in Grand Junction, CO, is $\frac{7}{10}$ inch in November, and $\frac{3}{5}$ inch in December. What is the total average precipitation?

Step 1 Find the LCD of the fractions.

$\frac{7}{10} + \frac{3}{5}$ (LCD, 10)

Step 2 Rename each numerator and each denominator with the LCD.

$7 \times 1 = 7, \quad 10 \times 1 = 10$

$3 \times 2 = 6, \quad 5 \times 2 = 10$

Step 3 Add the numerators.

$\frac{7}{10} + \frac{6}{10} = \frac{(7 + 6)}{10} = \frac{13}{10}$

$\frac{13}{10}$ inches total precipitation, or $1\frac{3}{10}$ inches.

Practice Problem On an electric bill, about $\frac{1}{8}$ of the energy is from solar energy and about $\frac{1}{10}$ is from wind power. How much of the total bill is from solar energy and wind power combined?

Example 3 In your body, $\frac{7}{10}$ of your muscle contractions are involuntary (cardiac and smooth muscle tissue). Smooth muscle makes $\frac{3}{15}$ of your muscle contractions. How many of your muscle contractions are made by cardiac muscle?

Step 1 Find the LCD of the fractions.

$\frac{7}{10} - \frac{3}{15}$ (LCD, 30)

Step 2 Rename each numerator and each denominator with the LCD.

$7 \times 3 = 21, \quad 10 \times 3 = 30$

$3 \times 2 = 6, \quad 15 \times 2 = 30$

Step 3 Subtract the numerators.

$\frac{21}{30} - \frac{6}{30} = \frac{(21 - 6)}{30} = \frac{15}{30}$

Step 4 Find the GCF.

$\frac{15}{30}$ (GCF, 15)

$\frac{1}{2}$

$\frac{1}{2}$ of all muscle contractions are cardiac muscle.

Example 4 Tony wants to make cookies that call for $\frac{3}{4}$ of a cup of flour, but he only has $\frac{1}{3}$ of a cup. How much more flour does he need?

Step 1 Find the LCD of the fractions.

$\frac{3}{4} - \frac{1}{3}$ (LCD, 12)

Step 2 Rename each numerator and each denominator with the LCD.

$3 \times 3 = 9, \quad 4 \times 3 = 12$

$1 \times 4 = 4, \quad 3 \times 4 = 12$

Step 3 Subtract the numerators.

$\frac{9}{12} - \frac{4}{12} = \frac{(9 - 4)}{12} = \frac{5}{12}$

$\frac{5}{12}$ of a cup of flour.

Practice Problem Using the information provided to you in Example 3 above, determine how many muscle contractions are voluntary (skeletal muscle).

Multiply Fractions To multiply with fractions, multiply the numerators and multiply the denominators. Find the simplest form if necessary.

Example Multiply $\frac{3}{5}$ by $\frac{1}{3}$.

Step 1 Multiply the numerators and denominators.

$$\frac{3}{5} \times \frac{1}{3} = \frac{(3 \times 1)}{(5 \times 3)} = \frac{3}{15}$$

Step 2 Find the GCF.

$$\frac{3}{15} \quad (\text{GCF, 3})$$

Step 3 Divide the numerator and denominator by the GCF.

$$\frac{3}{3} = 1, \quad \frac{15}{3} = 5$$

$$\frac{1}{5}$$

$\frac{3}{5}$ multiplied by $\frac{1}{3}$ is $\frac{1}{5}$.

Practice Problem Multiply $\frac{3}{14}$ by $\frac{5}{16}$.

Find a Reciprocal Two numbers whose product is 1 are called multiplicative inverses, or reciprocals.

Example Find the reciprocal of $\frac{3}{8}$.

Step 1 Inverse the fraction by putting the denominator on top and the numerator on the bottom.

$$\frac{8}{3}$$

The reciprocal of $\frac{3}{8}$ is $\frac{8}{3}$.

Practice Problem Find the reciprocal of $\frac{4}{9}$.

Divide Fractions To divide one fraction by another fraction, multiply the dividend by the reciprocal of the divisor. Find the simplest form if necessary.

Example 1 Divide $\frac{1}{9}$ by $\frac{1}{3}$.

Step 1 Find the reciprocal of the divisor.

The reciprocal of $\frac{1}{3}$ is $\frac{3}{1}$.

Step 2 Multiply the dividend by the reciprocal of the divisor.

$$\frac{\frac{1}{9}}{\frac{1}{3}} = \frac{1}{9} \times \frac{3}{1} = \frac{(1 \times 3)}{(9 \times 1)} = \frac{3}{9}$$

Step 3 Find the GCF.

$$\frac{3}{9} \quad (\text{GCF, 3})$$

Step 4 Divide the numerator and denominator by the GCF.

$$\frac{3}{3} = 1, \quad \frac{9}{3} = 3$$

$$\frac{1}{3}$$

$\frac{1}{9}$ divided by $\frac{1}{3}$ is $\frac{1}{3}$.

Example 2 Divide $\frac{3}{5}$ by $\frac{1}{4}$.

Step 1 Find the reciprocal of the divisor.

The reciprocal of $\frac{1}{4}$ is $\frac{4}{1}$.

Step 2 Multiply the dividend by the reciprocal of the divisor.

$$\frac{\frac{3}{5}}{\frac{1}{4}} = \frac{3}{5} \times \frac{4}{1} = \frac{(3 \times 4)}{(5 \times 1)} = \frac{12}{5}$$

$\frac{3}{5}$ divided by $\frac{1}{4}$ is $\frac{12}{5}$ or $2\frac{2}{5}$.

Practice Problem Divide $\frac{3}{11}$ by $\frac{7}{10}$.

Use Ratios

When you compare two numbers by division, you are using a ratio. Ratios can be written 3 to 5, 3:5, or $\frac{3}{5}$. Ratios, like fractions, also can be written in simplest form.

Ratios can represent probabilities, also called odds. This is a ratio that compares the number of ways a certain outcome occurs to the number of outcomes. For example, if you flip a coin 100 times, what are the odds that it will come up heads? There are two possible outcomes, heads or tails, so the odds of coming up heads are 50:100. Another way to say this is that 50 out of 100 times the coin will come up heads. In its simplest form, the ratio is 1:2.

Example 1 A chemical solution contains 40 g of salt and 64 g of baking soda. What is the ratio of salt to baking soda as a fraction in simplest form?

Step 1 Write the ratio as a fraction.
$$\frac{salt}{baking\ soda} = \frac{40}{64}$$

Step 2 Express the fraction in simplest form.
The GCF of 40 and 64 is 8.
$$\frac{40}{64} = \frac{40 \div 8}{64 \div 8} = \frac{5}{8}$$

The ratio of salt to baking soda in the sample is 5:8.

Example 2 Sean rolls a 6-sided die 6 times. What are the odds that the side with a 3 will show?

Step 1 Write the ratio as a fraction.
$$\frac{number\ of\ sides\ with\ a\ 3}{number\ of\ sides} = \frac{1}{6}$$

Step 2 Multiply by the number of attempts.
$$\frac{1}{6} \times 6\ attempts = \frac{6}{6}\ attempts = 1\ attempt$$

1 attempt out of 6 will show a 3.

Practice Problem Two metal rods measure 100 cm and 144 cm in length. What is the ratio of their lengths in simplest form?

Use Decimals

A fraction with a denominator that is a power of ten can be written as a decimal. For example, 0.27 means $\frac{27}{100}$. The decimal point separates the ones place from the tenths place.

Any fraction can be written as a decimal using division. For example, the fraction $\frac{5}{8}$ can be written as a decimal by dividing 5 by 8. Written as a decimal, it is 0.625.

Add or Subtract Decimals When adding and subtracting decimals, line up the decimal points before carrying out the operation.

Example 1 Find the sum of 47.68 and 7.80.

Step 1 Line up the decimal places when you write the numbers.
$$47.68$$
$$+\ 7.80$$

Step 2 Add the decimals.
$$47.68$$
$$+\ 7.80$$
$$\overline{55.48}$$

The sum of 47.68 and 7.80 is 55.48.

Example 2 Find the difference of 42.17 and 15.85.

Step 1 Line up the decimal places when you write the number.
$$42.17$$
$$-15.85$$

Step 2 Subtract the decimals.
$$42.17$$
$$-15.85$$
$$\overline{26.32}$$

The difference of 42.17 and 15.85 is 26.32.

Practice Problem Find the sum of 1.245 and 3.842.

Multiply Decimals To multiply decimals, multiply the numbers like any other number, ignoring the decimal point. Count the decimal places in each factor. The product will have the same number of decimal places as the sum of the decimal places in the factors.

Example Multiply 2.4 by 5.9.

Step 1 Multiply the factors like two whole numbers.
$24 \times 59 = 1416$

Step 2 Find the sum of the number of decimal places in the factors. Each factor has one decimal place, for a sum of two decimal places.

Step 3 The product will have two decimal places.
14.16

The product of 2.4 and 5.9 is 14.16.

Practice Problem Multiply 4.6 by 2.2.

Divide Decimals When dividing decimals, change the divisor to a whole number. To do this, multiply both the divisor and the dividend by the same power of ten. Then place the decimal point in the quotient directly above the decimal point in the dividend. Then divide as you do with whole numbers.

Example Divide 8.84 by 3.4.

Step 1 Multiply both factors by 10.
$3.4 \times 10 = 34, 8.84 \times 10 = 88.4$

Step 2 Divide 88.4 by 34.

```
      2.6
34)88.4
   -68
    204
   -204
      0
```

8.84 divided by 3.4 is 2.6.

Practice Problem Divide 75.6 by 3.6.

Use Proportions

An equation that shows that two ratios are equivalent is a proportion. The ratios $\frac{2}{4}$ and $\frac{5}{10}$ are equivalent, so they can be written as $\frac{2}{4} = \frac{5}{10}$. This equation is a proportion.

When two ratios form a proportion, the cross products are equal. To find the cross products in the proportion $\frac{2}{4} = \frac{5}{10}$, multiply the 2 and the 10, and the 4 and the 5. Therefore $2 \times 10 = 4 \times 5$, or $20 = 20$.

Because you know that both proportions are equal, you can use cross products to find a missing term in a proportion. This is known as solving the proportion.

Example The heights of a tree and a pole are proportional to the lengths of their shadows. The tree casts a shadow of 24 m when a 6-m pole casts a shadow of 4 m. What is the height of the tree?

Step 1 Write a proportion.
$$\frac{\text{height of tree}}{\text{height of pole}} = \frac{\text{length of tree's shadow}}{\text{length of pole's shadow}}$$

Step 2 Substitute the known values into the proportion. Let h represent the unknown value, the height of the tree.
$$\frac{h}{6} = \frac{24}{4}$$

Step 3 Find the cross products.
$h \times 4 = 6 \times 24$

Step 4 Simplify the equation.
$4h = 144$

Step 5 Divide each side by 4.
$$\frac{4h}{4} = \frac{144}{4}$$
$h = 36$

The height of the tree is 36 m.

Practice Problem The ratios of the weights of two objects on the Moon and on Earth are in proportion. A rock weighing 3 N on the Moon weighs 18 N on Earth. How much would a rock that weighs 5 N on the Moon weigh on Earth?

Use Percentages

The word *percent* means "out of one hundred." It is a ratio that compares a number to 100. Suppose you read that 77 percent of the Earth's surface is covered by water. That is the same as reading that the fraction of the Earth's surface covered by water is $\frac{77}{100}$. To express a fraction as a percent, first find the equivalent decimal for the fraction. Then, multiply the decimal by 100 and add the percent symbol.

Example Express $\frac{13}{20}$ as a percent.

Step 1 Find the equivalent decimal for the fraction.

$$
\begin{array}{r}
0.65 \\
20\overline{)13.00} \\
\underline{12\ 0} \\
1\ 00 \\
\underline{1\ 00} \\
0
\end{array}
$$

Step 2 Rewrite the fraction $\frac{13}{20}$ as 0.65.

Step 3 Multiply 0.65 by 100 and add the % sign.
$0.65 \times 100 = 65 = 65\%$

So, $\frac{13}{20} = 65\%$.

This also can be solved as a proportion.

Example Express $\frac{13}{20}$ as a percent.

Step 1 Write a proportion.
$$\frac{13}{20} = \frac{x}{100}$$

Step 2 Find the cross products.
$1300 = 20x$

Step 3 Divide each side by 20.
$$\frac{1300}{20} = \frac{20x}{20}$$
$65\% = x$

Practice Problem In one year, 73 of 365 days were rainy in one city. What percent of the days in that city were rainy?

Solve One-Step Equations

A statement that two things are equal is an equation. For example, $A = B$ is an equation that states that A is equal to B.

An equation is solved when a variable is replaced with a value that makes both sides of the equation equal. To make both sides equal the inverse operation is used. Addition and subtraction are inverses, and multiplication and division are inverses.

Example 1 Solve the equation $x - 10 = 35$.

Step 1 Find the solution by adding 10 to each side of the equation.
$x - 10 = 35$
$x - 10 + 10 = 35 + 10$
$x = 45$

Step 2 Check the solution.
$x - 10 = 35$
$45 - 10 = 35$
$35 = 35$

Both sides of the equation are equal, so $x = 45$.

Example 2 In the formula $a = bc$, find the value of c if $a = 20$ and $b = 2$.

Step 1 Rearrange the formula so the unknown value is by itself on one side of the equation by dividing both sides by b.

$a = bc$
$\frac{a}{b} = \frac{bc}{b}$
$\frac{a}{b} = c$

Step 2 Replace the variables a and b with the values that are given.

$\frac{a}{b} = c$
$\frac{20}{2} = c$
$10 = c$

Step 3 Check the solution.

$a = bc$
$20 = 2 \times 10$
$20 = 20$

Both sides of the equation are equal, so $c = 10$ is the solution when $a = 20$ and $b = 2$.

Practice Problem In the formula $h = gd$, find the value of d if $g = 12.3$ and $h = 17.4$.

Use Statistics

The branch of mathematics that deals with collecting, analyzing, and presenting data is statistics. In statistics, there are three common ways to summarize data with a single number—the mean, the median, and the mode.

The **mean** of a set of data is the arithmetic average. It is found by adding the numbers in the data set and dividing by the number of items in the set.

The **median** is the middle number in a set of data when the data are arranged in numerical order. If there were an even number of data points, the median would be the mean of the two middle numbers.

The **mode** of a set of data is the number or item that appears most often.

Another number that often is used to describe a set of data is the range. The **range** is the difference between the largest number and the smallest number in a set of data.

A **frequency table** shows how many times each piece of data occurs, usually in a survey. **Table 2** below shows the results of a student survey on favorite color.

Table 2 Student Color Choice

Color	Tally	Frequency
red	||||	4
blue	₩₩	5
black	||	2
green	|||	3
purple	₩₩ ||	7
yellow	₩₩ |	6

Based on the frequency table data, which color is the favorite?

Example The speeds (in m/s) for a race car during five different time trials are 39, 37, 44, 36, and 44.

To find the mean:

Step 1 Find the sum of the numbers.
$39 + 37 + 44 + 36 + 44 = 200$

Step 2 Divide the sum by the number of items, which is 5.
$200 \div 5 = 40$

The mean is 40 m/s.

To find the median:

Step 1 Arrange the measures from least to greatest.
36, 37, 39, 44, 44

Step 2 Determine the middle measure.
36, 37, 39, 44, 44

The median is 39 m/s.

To find the mode:

Step 1 Group the numbers that are the same together.
44, 44, 36, 37, 39

Step 2 Determine the number that occurs most in the set.
44, 44, 36, 37, 39

The mode is 44 m/s.

To find the range:

Step 1 Arrange the measures from largest to smallest.
44, 44, 39, 37, 36

Step 2 Determine the largest and smallest measures in the set.
44, 44, 39, 37, 36

Step 3 Find the difference between the largest and smallest measures.
$44 - 36 = 8$

The range is 8 m/s.

Practice Problem Find the mean, median, mode, and range for the data set 8, 4, 12, 8, 11, 14, 16.

Use Geometry

The branch of mathematics that deals with the measurement, properties, and relationships of points, lines, angles, surfaces, and solids is called geometry.

Perimeter The **perimeter** (P) is the distance around a geometric figure. To find the perimeter of a rectangle, add the length and width and multiply that sum by two, or $2(l + w)$. To find perimeters of irregular figures, add the length of the sides.

Example 1 Find the perimeter of a rectangle that is 3 m long and 5 m wide.

Step 1 You know that the perimeter is 2 times the sum of the width and length.
$$P = 2(3 \text{ m} + 5 \text{ m})$$

Step 2 Find the sum of the width and length.
$$P = 2(8 \text{ m})$$

Step 3 Multiply by 2.
$$P = 16 \text{ m}$$

The perimeter is 16 m.

Example 2 Find the perimeter of a shape with sides measuring 2 cm, 5 cm, 6 cm, 3 cm.

Step 1 You know that the perimeter is the sum of all the sides.
$$P = 2 + 5 + 6 + 3$$

Step 2 Find the sum of the sides.
$$P = 2 + 5 + 6 + 3$$
$$P = 16$$

The perimeter is 16 cm.

Practice Problem Find the perimeter of a rectangle with a length of 18 m and a width of 7 m.

Practice Problem Find the perimeter of a triangle measuring 1.6 cm by 2.4 cm by 2.4 cm.

Area of a Rectangle The **area** (A) is the number of square units needed to cover a surface. To find the area of a rectangle, multiply the length times the width, or $l \times w$. When finding area, the units also are multiplied. Area is given in square units.

Example Find the area of a rectangle with a length of 1 cm and a width of 10 cm.

Step 1 You know that the area is the length multiplied by the width.
$$A = (1 \text{ cm} \times 10 \text{ cm})$$

Step 2 Multiply the length by the width. Also multiply the units.
$$A = 10 \text{ cm}^2$$

The area is 10 cm².

Practice Problem Find the area of a square whose sides measure 4 m.

Area of a Triangle To find the area of a triangle, use the formula:

$$A = \frac{1}{2}(\text{base} \times \text{height})$$

The base of a triangle can be any of its sides. The height is the perpendicular distance from a base to the opposite endpoint, or vertex.

Example Find the area of a triangle with a base of 18 m and a height of 7 m.

Step 1 You know that the area is $\frac{1}{2}$ the base times the height.
$$A = \frac{1}{2}(18 \text{ m} \times 7 \text{ m})$$

Step 2 Multiply $\frac{1}{2}$ by the product of 18×7. Multiply the units.
$$A = \frac{1}{2}(126 \text{ m}^2)$$
$$A = 63 \text{ m}^2$$

The area is 63 m².

Practice Problem Find the area of a triangle with a base of 27 cm and a height of 17 cm.

Circumference of a Circle The **diameter** (*d*) of a circle is the distance across the circle through its center, and the **radius** (*r*) is the distance from the center to any point on the circle. The radius is half of the diameter. The distance around the circle is called the **circumference** (C). The formula for finding the circumference is:

$$C = 2\pi r \ \text{ or } \ C = \pi d$$

The circumference divided by the diameter is always equal to 3.1415926... This nonterminating and nonrepeating number is represented by the Greek letter π (pi). An approximation often used for π is 3.14.

Example 1 Find the circumference of a circle with a radius of 3 m.

Step 1 You know the formula for the circumference is 2 times the radius times π.
$$C = 2\pi(3)$$

Step 2 Multiply 2 times the radius.
$$C = 6\pi$$

Step 3 Multiply by π.
$$C = 19 \text{ m}$$

The circumference is 19 m.

Example 2 Find the circumference of a circle with a diameter of 24.0 cm.

Step 1 You know the formula for the circumference is the diameter times π.
$$C = \pi(24.0)$$

Step 2 Multiply the diameter by π.
$$C = 75.4 \text{ cm}$$

The circumference is 75.4 cm.

Practice Problem Find the circumference of a circle with a radius of 19 cm.

Area of a Circle The formula for the area of a circle is:
$$A = \pi r^2$$

Example 1 Find the area of a circle with a radius of 4.0 cm.

Step 1 $A = \pi(4.0)^2$

Step 2 Find the square of the radius.
$$A = 16\pi$$

Step 3 Multiply the square of the radius by π.
$$A = 50 \text{ cm}^2$$

The area of the circle is 50 cm^2.

Example 2 Find the area of a circle with a radius of 225 m.

Step 1 $A = \pi(225)^2$

Step 2 Find the square of the radius.
$$A = 50625\pi$$

Step 3 Multiply the square of the radius by π.
$$A = 158962.5$$

The area of the circle is 158,962 m^2.

Example 3 Find the area of a circle whose diameter is 20.0 mm.

Step 1 You know the formula for the area of a circle is the square of the radius times π, and that the radius is half of the diameter.
$$A = \pi\left(\frac{20.0}{2}\right)^2$$

Step 2 Find the radius.
$$A = \pi(10.0)^2$$

Step 3 Find the square of the radius.
$$A = 100\pi$$

Step 4 Multiply the square of the radius by π.
$$A = 314 \text{ mm}^2$$

The area is 314 mm^2.

Practice Problem Find the area of a circle with a radius of 16 m.

Volume The measure of space occupied by a solid is the **volume** (V). To find the volume of a rectangular solid multiply the length times width times height, or $V = l \times w \times h$. It is measured in cubic units, such as cubic centimeters (cm^3).

Example Find the volume of a rectangular solid with a length of 2.0 m, a width of 4.0 m, and a height of 3.0 m.

Step 1 You know the formula for volume is the length times the width times the height.

$V = 2.0 \text{ m} \times 4.0 \text{ m} \times 3.0 \text{ m}$

Step 2 Multiply the length times the width times the height.

$V = 24 \text{ m}^3$

The volume is 24 m³.

Practice Problem Find the volume of a rectangular solid that is 8 m long, 4 m wide, and 4 m high.

To find the volume of other solids, multiply the area of the base times the height.

Example 1 Find the volume of a solid that has a triangular base with a length of 8.0 m and a height of 7.0 m. The height of the entire solid is 15.0 m.

Step 1 You know that the base is a triangle, and the area of a triangle is $\frac{1}{2}$ the base times the height, and the volume is the area of the base times the height.

$V = \left[\frac{1}{2}(b \times h)\right] \times 15$

Step 2 Find the area of the base.

$V = \left[\frac{1}{2}(8 \times 7)\right] \times 15$

$V = \left(\frac{1}{2} \times 56\right) \times 15$

Step 3 Multiply the area of the base by the height of the solid.

$V = 28 \times 15$

$V = 420 \text{ m}^3$

The volume is 420 m³.

Example 2 Find the volume of a cylinder that has a base with a radius of 12.0 cm, and a height of 21.0 cm.

Step 1 You know that the base is a circle, and the area of a circle is the square of the radius times π, and the volume is the area of the base times the height.

$V = (\pi r^2) \times 21$

$V = (\pi 12^2) \times 21$

Step 2 Find the area of the base.

$V = 144\pi \times 21$

$V = 452 \times 21$

Step 3 Multiply the area of the base by the height of the solid.

$V = 9490 \text{ cm}^3$

The volume is 9490 cm³.

Example 3 Find the volume of a cylinder that has a diameter of 15 mm and a height of 4.8 mm.

Step 1 You know that the base is a circle with an area equal to the square of the radius times π. The radius is one-half the diameter. The volume is the area of the base times the height.

$V = (\pi r^2) \times 4.8$

$V = \left[\pi\left(\frac{1}{2} \times 15\right)^2\right] \times 4.8$

$V = (\pi 7.5^2) \times 4.8$

Step 2 Find the area of the base.

$V = 56.25\pi \times 4.8$

$V = 176.63 \times 4.8$

Step 3 Multiply the area of the base by the height of the solid.

$V = 847.8$

The volume is 847.8 mm³.

Practice Problem Find the volume of a cylinder with a diameter of 7 cm in the base and a height of 16 cm.

Science Applications

Measure in SI

The metric system of measurement was developed in 1795. A modern form of the metric system, called the International System (SI), was adopted in 1960 and provides the standard measurements that all scientists around the world can understand.

The SI system is convenient because unit sizes vary by powers of 10. Prefixes are used to name units. Look at **Table 3** for some common SI prefixes and their meanings.

Table 3 Common SI Prefixes			
Prefix	**Symbol**	**Meaning**	
kilo-	k	1,000	thousand
hecto-	h	100	hundred
deka-	da	10	ten
deci-	d	0.1	tenth
centi-	c	0.01	hundredth
milli-	m	0.001	thousandth

Example How many grams equal one kilogram?

Step 1 Find the prefix *kilo* in **Table 3.**

Step 2 Using **Table 3,** determine the meaning of *kilo.* According to the table, it means 1,000. When the prefix *kilo* is added to a unit, it means that there are 1,000 of the units in a "*kilo*unit."

Step 3 Apply the prefix to the units in the question. The units in the question are grams. There are 1,000 grams in a kilogram.

Practice Problem Is a milligram larger or smaller than a gram? How many of the smaller units equal one larger unit? What fraction of the larger unit does one smaller unit represent?

Dimensional Analysis

Convert SI Units In science, quantities such as length, mass, and time sometimes are measured using different units. A process called dimensional analysis can be used to change one unit of measure to another. This process involves multiplying your starting quantity and units by one or more conversion factors. A conversion factor is a ratio equal to one and can be made from any two equal quantities with different units. If 1,000 mL equal 1 L then two ratios can be made.

$$\frac{1{,}000 \text{ mL}}{1 \text{ L}} = \frac{1 \text{ L}}{1{,}000 \text{ mL}} = 1$$

One can covert between units in the SI system by using the equivalents in **Table 3** to make conversion factors.

Example 1 How many cm are in 4 m?

Step 1 Write conversion factors for the units given. From **Table 3,** you know that 100 cm = 1 m. The conversion factors are

$$\frac{100 \text{ cm}}{1 \text{ m}} \quad and \quad \frac{1 \text{ m}}{100 \text{ cm}}$$

Step 2 Decide which conversion factor to use. Select the factor that has the units you are converting from (m) in the denominator and the units you are converting to (cm) in the numerator.

$$\frac{100 \text{ cm}}{1 \text{ m}}$$

Step 3 Multiply the starting quantity and units by the conversion factor. Cancel the starting units with the units in the denominator. There are 400 cm in 4 m.

$$4 \text{ m} \times \frac{100 \text{ cm}}{1 \text{ m}} = 400 \text{ cm}$$

Practice Problem How many milligrams are in one kilogram? (Hint: You will need to use two conversion factors from **Table 3.**)

Table 4 Unit System Equivalents

Type of Measurement	Equivalent
Length	1 in = 2.54 cm 1 yd = 0.91 m 1 mi = 1.61 km
Mass and Weight*	1 oz = 28.35 g 1 lb = 0.45 kg 1 ton (short) = 0.91 tonnes (metric tons) 1 lb = 4.45 N
Volume	$1\ in^3 = 16.39\ cm^3$ 1 qt = 0.95 L 1 gal = 3.78 L
Area	$1\ in^2 = 6.45\ cm^2$ $1\ yd^2 = 0.83\ m^2$ $1\ mi^2 = 2.59\ km^2$ 1 acre = 0.40 hectares
Temperature	$°C = \dfrac{(°F - 32)}{1.8}$ $K = °C + 273$

*Weight is measured in standard Earth gravity.

Convert Between Unit Systems Table 4 gives a list of equivalents that can be used to convert between English and SI units.

Example   If a meterstick has a length of 100 cm, how long is the meterstick in inches?

Step 1 Write the conversion factors for the units given. From **Table 4,** 1 in = 2.54 cm.

$$\frac{1\ in}{2.54\ cm}\ \ and\ \ \frac{2.54\ cm}{1\ in}$$

Step 2 Determine which conversion factor to use. You are converting from cm to in. Use the conversion factor with cm on the bottom.

$$\frac{1\ in}{2.54\ cm}$$

Step 3 Multiply the starting quantity and units by the conversion factor. Cancel the starting units with the units in the denominator. Round your answer based on the number of significant figures in the conversion factor.

$$100\ \cancel{cm} \times \frac{1\ in}{2.54\ \cancel{cm}} = 39.37\ in$$

The meterstick is 39.4 in long.

Practice Problem   A book has a mass of 5 lbs. What is the mass of the book in kg?

Practice Problem   Use the equivalent for in and cm (1 in = 2.54 cm) to show how $1\ in^3 = 16.39\ cm^3$.

Precision and Significant Digits

When you make a measurement, the value you record depends on the precision of the measuring instrument. This precision is represented by the number of significant digits recorded in the measurement. When counting the number of significant digits, all digits are counted except zeros at the end of a number with no decimal point such as 2,050, and zeros at the beginning of a decimal such as 0.03020. When adding or subtracting numbers with different precision, round the answer to the smallest number of decimal places of any number in the sum or difference. When multiplying or dividing, the answer is rounded to the smallest number of significant digits of any number being multiplied or divided.

Example The lengths 5.28 and 5.2 are measured in meters. Find the sum of these lengths and record your answer using the correct number of significant digits.

Step 1 Find the sum.

5.28 m	2 digits after the decimal
+ 5.2 m	1 digit after the decimal
10.48 m	

Step 2 Round to one digit after the decimal because the least number of digits after the decimal of the numbers being added is 1.

The sum is 10.5 m.

Practice Problem How many significant digits are in the measurement 7,071,301 m? How many significant digits are in the measurement 0.003010 g?

Practice Problem Multiply 5.28 and 5.2 using the rule for multiplying and dividing. Record the answer using the correct number of significant digits.

Scientific Notation

Many times numbers used in science are very small or very large. Because these numbers are difficult to work with scientists use scientific notation. To write numbers in scientific notation, move the decimal point until only one non-zero digit remains on the left. Then count the number of places you moved the decimal point and use that number as a power of ten. For example, the average distance from the Sun to Mars is 227,800,000,000 m. In scientific notation, this distance is 2.278×10^{11} m. Because you moved the decimal point to the left, the number is a positive power of ten.

The mass of an electron is about 0.000 000 000 000 000 000 000 000 000 000 911 kg. Expressed in scientific notation, this mass is 9.11×10^{-31} kg. Because the decimal point was moved to the right, the number is a negative power of ten.

Example Earth is 149,600,000 km from the Sun. Express this in scientific notation.

Step 1 Move the decimal point until one non-zero digit remains on the left.
1.496 000 00

Step 2 Count the number of decimal places you have moved. In this case, eight.

Step 3 Show that number as a power of ten, 10^8.

The Earth is 1.496×10^8 km from the Sun.

Practice Problem How many significant digits are in 149,600,000 km? How many significant digits are in 1.496×10^8 km?

Practice Problem Parts used in a high performance car must be measured to 7×10^{-6} m. Express this number as a decimal.

Practice Problem A CD is spinning at 539 revolutions per minute. Express this number in scientific notation.

Make and Use Graphs

Data in tables can be displayed in a graph—a visual representation of data. Common graph types include line graphs, bar graphs, and circle graphs.

Line Graph A line graph shows a relationship between two variables that change continuously. The independent variable is changed and is plotted on the *x*-axis. The dependent variable is observed, and is plotted on the *y*-axis.

Example Draw a line graph of the data below from a cyclist in a long-distance race.

Table 5 Bicycle Race Data	
Time (h)	**Distance (km)**
0	0
1	8
2	16
3	24
4	32
5	40

Step 1 Determine the *x*-axis and *y*-axis variables. Time varies independently of distance and is plotted on the *x*-axis. Distance is dependent on time and is plotted on the *y*-axis.

Step 2 Determine the scale of each axis. The *x*-axis data ranges from 0 to 5. The *y*-axis data ranges from 0 to 40.

Step 3 Using graph paper, draw and label the axes. Include units in the labels.

Step 4 Draw a point at the intersection of the time value on the *x*-axis and corresponding distance value on the *y*-axis. Connect the points and label the graph with a title, as shown in **Figure 20.**

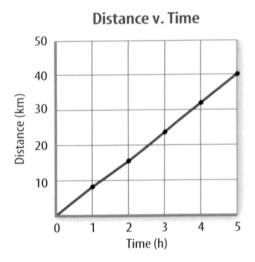

Distance v. Time

Figure 20 This line graph shows the relationship between distance and time during a bicycle ride.

Practice Problem A puppy's shoulder height is measured during the first year of her life. The following measurements were collected: (3 mo, 52 cm), (6 mo, 72 cm), (9 mo, 83 cm), (12 mo, 86 cm). Graph this data.

Find a Slope The slope of a straight line is the ratio of the vertical change, rise, to the horizontal change, run.

$$\text{Slope} = \frac{\text{vertical change (rise)}}{\text{horizontal change (run)}} = \frac{\text{change in } y}{\text{change in } x}$$

Example Find the slope of the graph in **Figure 20.**

Step 1 You know that the slope is the change in *y* divided by the change in *x*.

$$\text{Slope} = \frac{\text{change in } y}{\text{change in } x}$$

Step 2 Determine the data points you will be using. For a straight line, choose the two sets of points that are the farthest apart.

$$\text{Slope} = \frac{(40-0) \text{ km}}{(5-0) \text{ hr}}$$

Step 3 Find the change in *y* and *x*.

$$\text{Slope} = \frac{40 \text{ km}}{5 \text{h}}$$

Step 4 Divide the change in *y* by the change in *x*.

$$\text{Slope} = \frac{8 \text{ km}}{\text{h}}$$

The slope of the graph is 8 km/h.

Bar Graph To compare data that does not change continuously you might choose a bar graph. A bar graph uses bars to show the relationships between variables. The x-axis variable is divided into parts. The parts can be numbers such as years, or a category such as a type of animal. The y-axis is a number and increases continuously along the axis.

Example A recycling center collects 4.0 kg of aluminum on Monday, 1.0 kg on Wednesday, and 2.0 kg on Friday. Create a bar graph of this data.

Step 1 Select the x-axis and y-axis variables. The measured numbers (the masses of aluminum) should be placed on the y-axis. The variable divided into parts (collection days) is placed on the x-axis.

Step 2 Create a graph grid like you would for a line graph. Include labels and units.

Step 3 For each measured number, draw a vertical bar above the x-axis value up to the y-axis value. For the first data point, draw a vertical bar above Monday up to 4.0 kg.

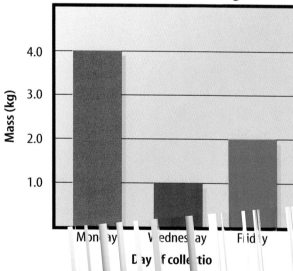

Aluminum Collected During Week

Practice Problem Draw a bar graph of the gases in air: 78% nitrogen, 21% oxygen, 1% other gases.

Circle Graph To display data as parts of a whole, you might use a circle graph. A circle graph is a circle divided into sections that represent the relative size of each piece of data. The entire circle represents 100%, half represents 50%, and so on.

Example Air is made up of 78% nitrogen, 21% oxygen, and 1% other gases. Display the composition of air in a circle graph.

Step 1 Multiply each percent by 360° and divide by 100 to find the angle of each section in the circle.

$$78\% \times \frac{360°}{100} = 280.8°$$

$$21\% \times \frac{360°}{100} = 75.6°$$

$$1\% \times \frac{360°}{100} = 3.6°$$

Step 2 Use a compass to draw a circle and to mark the center of the circle. Draw a straight line from the center to the edge of the circle.

Step 3 Use a protractor and the angles you calculated to divide the circle into parts. Place the center of the protractor over the center of the circle and line the base of the protractor over the straight line.

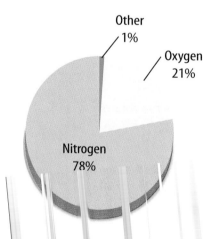

Practice Problem Draw a circle graph to represent the amount of aluminum collected during the week shown in the bar graph to the left.

Physical Science Reference Tables

Standard Units

Symbol	Name	Quantity
m	meter	length
kg	kilogram	mass
Pa	pascal	pressure
K	kelvin	temperature
mol	mole	amount of a substance
J	joule	energy, work, quantity of heat
s	second	time
C	coulomb	electric charge
V	volt	electric potential
A	ampere	electric current
Ω	ohm	resistance

Wavelengths of Light in a Vacuum

Violet	$4.0 - 4.2 \times 10^{-7}$ m
Blue	$4.2 - 4.9 \times 10^{-7}$ m
Green	$4.9 - 5.7 \times 10^{-7}$ m
Yellow	$5.7 - 5.9 \times 10^{-7}$ m
Orange	$5.9 - 6.5 \times 10^{-7}$ m
Red	$6.5 - 7.0 \times 10^{-7}$ m

Physical Constants and Conversion Factors

Acceleration due to gravity	g	9.8 m/s/s or m/s^2
Avogadro's Number	N_A	6.02×10^{23} particles per mole
Electron charge	e	1.6×10^{-19} C
Electron rest mass	m_e	9.11×10^{-31} kg
Gravitation constant	G	6.67×10^{-11} N $\times$ m^2/kg^2
Mass-energy relationship		1 u (amu) = 9.3×10^2 MeV
Speed of light in a vacuum	c	3.00×10^8 m/s
Speed of sound at STP		331 m/s
Standard Pressure		1 atmosphere
		101.3 kPa
		760 Torr or mmHg
		14.7 lb/in.^{2}

The Index of Refraction for Common Substances
($\lambda = 5.9 \times 10^{-7}$ m)

Air	1.00
Alcohol	1.36
Canada Balsam	1.53
Corn Oil	1.47
Diamond	2.42
Glass, Crown	1.52
Glass, Flint	1.61
Glycerol	1.47
Lucite	1.50
Quartz, Fused	1.46
Water	1.33

Heat Constants

	Specific Heat (average) (kJ/kg $\times$ °C) (J/g $\times$ °C)	Melting Point (°C)	Boiling Point (°C)	Heat of Fusion (kJ/kg) (J/g)	Heat of Vaporization (kJ/kg) (J/g)
Alcohol (ethyl)	2.43 (liq.)	-117	79	109	855
Aluminum	0.90 (sol.)	660	2467	396	10500
Ammonia	4.71 (liq.)	-78	-33	332	1370
Copper	0.39 (sol.)	1083	2567	205	4790
Iron	0.45 (sol.)	1535	2750	267	6290
Lead	0.13 (sol.)	328	1740	25	866
Mercury	0.14 (liq.)	-39	357	11	295
Platinum	0.13 (sol.)	1772	3827	101	229
Silver	0.24 (sol.)	962	2212	105	2370
Tungsten	0.13 (sol.)	3410	5660	192	4350
Water (solid)	2.05 (sol.)	0	–	334	–
Water (liquid)	4.18 (liq.)	–	100	–	–
Water (vapor)	2.01 (gas)	–	–	–	2260
Zinc	0.39 (sol.)	420	907	113	1770

PERIODIC TABLE OF THE ELEMENTS

Columns of elements are called groups. Elements in the same group have similar chemical properties.

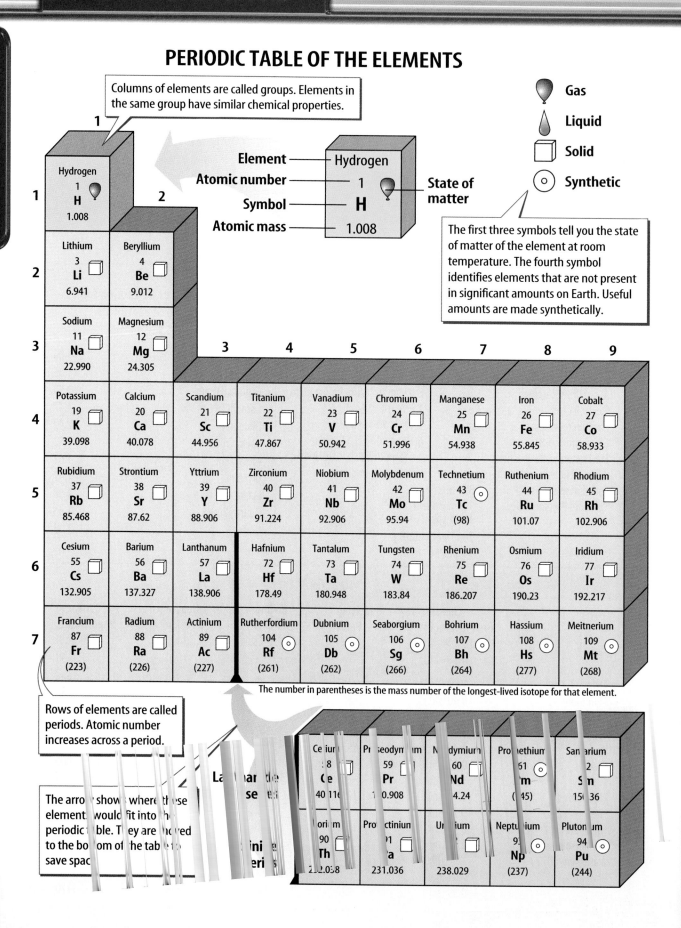

Gas

Liquid

Solid

Synthetic

Element — Hydrogen
Atomic number — 1
Symbol — H
Atomic mass — 1.008

State of matter

The first three symbols tell you the state of matter of the element at room temperature. The fourth symbol identifies elements that are not present in significant amounts on Earth. Useful amounts are made synthetically.

	1	2	3	4	5	6	7	8	9
1	Hydrogen 1 **H** 1.008								
2	Lithium 3 **Li** 6.941	Beryllium 4 **Be** 9.012							
3	Sodium 11 **Na** 22.990	Magnesium 12 **Mg** 24.305							
4	Potassium 19 **K** 39.098	Calcium 20 **Ca** 40.078	Scandium 21 **Sc** 44.956	Titanium 22 **Ti** 47.867	Vanadium 23 **V** 50.942	Chromium 24 **Cr** 51.996	Manganese 25 **Mn** 54.938	Iron 26 **Fe** 55.845	Cobalt 27 **Co** 58.933
5	Rubidium 37 **Rb** 85.468	Strontium 38 **Sr** 87.62	Yttrium 39 **Y** 88.906	Zirconium 40 **Zr** 91.224	Niobium 41 **Nb** 92.906	Molybdenum 42 **Mo** 95.94	Technetium 43 **Tc** (98)	Ruthenium 44 **Ru** 101.07	Rhodium 45 **Rh** 102.906
6	Cesium 55 **Cs** 132.905	Barium 56 **Ba** 137.327	Lanthanum 57 **La** 138.906	Hafnium 72 **Hf** 178.49	Tantalum 73 **Ta** 180.948	Tungsten 74 **W** 183.84	Rhenium 75 **Re** 186.207	Osmium 76 **Os** 190.23	Iridium 77 **Ir** 192.217
7	Francium 87 **Fr** (223)	Radium 88 **Ra** (226)	Actinium 89 **Ac** (227)	Rutherfordium 104 **Rf** (261)	Dubnium 105 **Db** (262)	Seaborgium 106 **Sg** (266)	Bohrium 107 **Bh** (264)	Hassium 108 **Hs** (277)	Meitnerium 109 **Mt** (268)

The number in parentheses is the mass number of the longest-lived isotope for that element.

Rows of elements are called periods. Atomic number increases across a period.

The arrow shows where these elements would fit into the periodic table. They are moved to the bottom of the table to save space.

Lanthanide series	Cerium 58 **Ce** 140.116	Praseodymium 59 **Pr** 140.908	Neodymium 60 **Nd** 144.24	Promethium 61 **Pm** (145)	Samarium 62 **Sm** 150.36
Actinide series	Thorium 90 **Th** 232.038	Protactinium 91 **Pa** 231.036	Uranium 92 **U** 238.029	Neptunium 93 **Np** (237)	Plutonium 94 **Pu** (244)

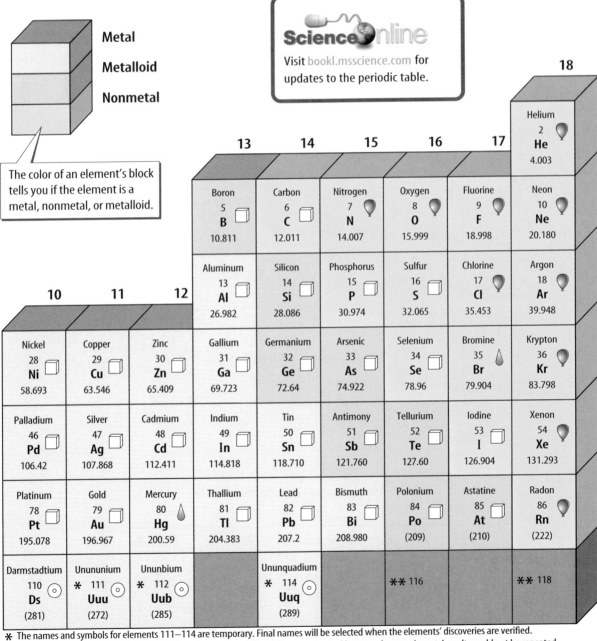

Metal
Metalloid
Nonmetal

The color of an element's block tells you if the element is a metal, nonmetal, or metalloid.

Science Online
Visit bookl.msscience.com for updates to the periodic table.

		18
		Helium 2 He 4.003

13	14	15	16	17	
Boron 5 B 10.811	Carbon 6 C 12.011	Nitrogen 7 N 14.007	Oxygen 8 O 15.999	Fluorine 9 F 18.998	Neon 10 Ne 20.180
Aluminum 13 Al 26.982	Silicon 14 Si 28.086	Phosphorus 15 P 30.974	Sulfur 16 S 32.065	Chlorine 17 Cl 35.453	Argon 18 Ar 39.948

10	11	12						
Nickel 28 Ni 58.693	Copper 29 Cu 63.546	Zinc 30 Zn 65.409	Gallium 31 Ga 69.723	Germanium 32 Ge 72.64	Arsenic 33 As 74.922	Selenium 34 Se 78.96	Bromine 35 Br 79.904	Krypton 36 Kr 83.798
Palladium 46 Pd 106.42	Silver 47 Ag 107.868	Cadmium 48 Cd 112.411	Indium 49 In 114.818	Tin 50 Sn 118.710	Antimony 51 Sb 121.760	Tellurium 52 Te 127.60	Iodine 53 I 126.904	Xenon 54 Xe 131.293
Platinum 78 Pt 195.078	Gold 79 Au 196.967	Mercury 80 Hg 200.59	Thallium 81 Tl 204.383	Lead 82 Pb 207.2	Bismuth 83 Bi 208.980	Polonium 84 Po (209)	Astatine 85 At (210)	Radon 86 Rn (222)
Darmstadtium 110 Ds (281)	Unununium * 111 Uuu (272)	Ununbium * 112 Uub (285)		Ununquadium * 114 Uuq (289)		** 116		** 118

* The names and symbols for elements 111–114 are temporary. Final names will be selected when the elements' discoveries are verified.

** Elements 116 and 118 were thought to have been created. The claim was retracted because the experimental results could not be repeated.

Europium 63 Eu 151.964	Gadolinium 64 Gd 157.25	Terbium 65 Tb 158.925	Dysprosium 66 Dy 162.500	Holmium 67 Ho 164.930	Erbium 68 Er 167.259	Thulium 69 Tm 168.934	Ytterbium 70 Yb 173.04	Lutetium 71 Lu 174.967
Americium 95 Am (243)	Curium 96 Cm (247)	Berkelium 97 Bk (247)	Californium 98 Cf (251)	Einsteinium 99 Es (252)	Fermium 100 Fm (257)	Mendelevium 101 Md (258)	Nobelium 102 No (259)	Lawrencium 103 Lr (262)

Standard Units

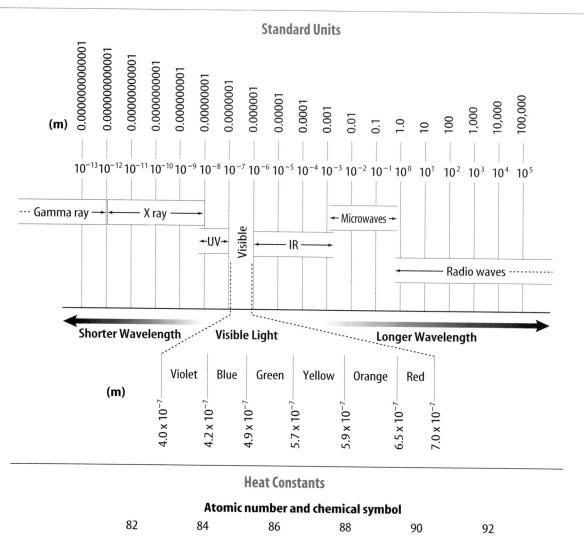

Heat Constants

Atomic number and chemical symbol

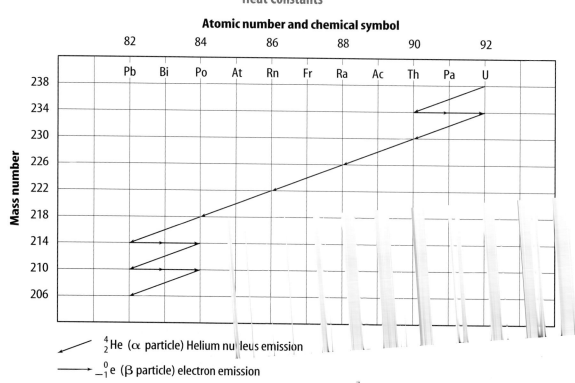

$^4_2$He (α particle) Helium nucleus emission

$^0_{-1}$e (β particle) electron emission

Cómo usar el glosario en español:
1. Busca el término en inglés que desees encontrar.
2. El término en español, junto con la definición, se encuentran en la columna de la derecha.

Pronunciation Key

Use the following key to help you sound out words in the glossary.

a	back (BAK)	ew	food (FEWD)
ay	day (DAY)	yoo	pure (PYOOR)
ah	father (FAH thur)	yew	few (FYEW)
ow	flower (FLOW ur)	uh	comma (CAH muh)
ar	car (CAR)	u (+ con)	rub (RUB)
e	less (LES)	sh	shelf (SHELF)
ee	leaf (LEEF)	ch	nature (NAY chur)
ih	trip (TRIHP)	g	gift (GIHFT)
i (i + con + e)	idea (i DEE uh)	j	gem (JEM)
oh	go (GOH)	ing	sing (SING)
aw	soft (SAWFT)	zh	vision (VIH zhun)
or	orbit (OR buht)	k	cake (KAYK)
oy	coin (COYN)	s	seed, cent (SEED, SENT)
oo	foot (FOOT)	z	zone, raise (ZOHN, RAYZ)

English   **Español**

A

acid: substance that releases H⁺ ions and produces hydronium ions when dissolved in water. (p. 78)

activation energy: minimum amount of energy needed to start a chemical reaction. (p. 47)

amino acids: building blocks of proteins; contain both an amino group and a carboxyl acid group replacing hydrogens on the same carbon atom. (p. 105)

amino (uh MEE noh) group: consists of one nitrogen atom covalently bonded to two hydrogen atoms; represented by the formula –NH₂. (p. 105)

aqueous (A kwee us): solution in which water is the solvent. (p. 70)

ácido: sustancia que libera iones H⁺ y produce iones de hidronio al ser disuelta en agua. (p. 78)

energía de activación: cantidad mínima de energía necesaria para iniciar una reacción química. (p. 47)

aminoácidos: bloques de construcción de las proteínas que contienen un grupo amino y un grupo ácido carboxilo reemplazando hidrógenos en el mismo átomo de carbono. (p. 105)

grupo amino: consiste en un átomo de nitrógeno unido por enlaces covalentes a dos átomos de hidrógeno; se lo representa con la fórmula –NH₂. (p. 105)

acuoso: solución en la cual el agua es el solvente. (p. 70)

B

base: substance that accepts H⁺ ions and produces hydroxide ions when dissolved in water. (p. 81)

base: sustancia que acepta los iones H⁺ y produce iones de hidróxido al ser disuelta en agua. (p. 81)

C

carbohydrates: organic compounds containing only carbon, hydrogen, and oxygen; starches, cellulose, glycogen, sugars. (p. 110)

carbohidratos: compuestos orgánicos que sólo contienen carbono, hidrógeno y oxígeno; ejemplos son los almidones, la celulosa, el glucógeno y los azúcares. (p. 110)

Glossary/Glosario

carboxyl (car BOK sul) group: consists of one carbon atom, two oxygen atoms, and one hydrogen atom; represented by the formula –COOH. (p. 105)

catalyst: substance that speeds up a chemical reaction but is not used up itself or permanently changed. (p. 51)

chemical bond: force that holds two atoms together. (p. 15)

chemical equation: shorthand form for writing what reactants are used and what products are formed in a chemical reaction; sometimes shows whether energy is produced or absorbed. (p. 38)

chemical formula: combination of chemical symbols and numbers that indicates which elements and how many atoms of each element are present in a molecule. (p. 24)

chemical reaction: process that produces chemical change, resulting in new substances that have properties different from those of the original substances. (p. 36)

cholesterol: a complex lipid that is present in foods that come from animals. (p. 115)

compound: pure substance that contains two or more elements. (p. 17)

concentration: describes how much solute is present in a solution compared to the amount of solvent. (pp. 49, 75)

covalent bond: chemical bond formed when atoms share electrons. (p. 19)

grupo carboxilo: consiste en un átomo de carbono, dos de oxígeno y uno de hidrógeno; se lo representa con la fórmula –COOH. (p. 105)

catalizador: sustancia que acelera una reacción química pero que ella misma ni se agota ni sufre cambios permanentes. (p. 51)

enlace químico: fuerza que mantiene a dos átomos unidos. (p. 15)

ecuación química: forma breve para representar los reactivos utilizados y los productos que se forman en una reacción química; algunas veces muestra si se produce o absorbe energía. (p. 38)

fórmula química: combinación de símbolos y números químicos que indican cuáles elementos y cuántos átomos de cada elemento están presentes en una molécula. (p. 24)

reacción química: proceso que produce cambios químicos que dan como resultado nuevas sustancias cuyas propiedades son diferentes a aquellas de las sustancias originales. (p. 36)

colesterol: lípido complejo presente en alimentos de origen animal. (p. 115)

compuesto: sustancia pura que contiene dos o más elementos. (p. 17)

concentración: describe la cantidad de soluto presente en una solución en relación con la cantidad de solvente. (pp. 49, 75)

enlace covalente: enlace químico que se forma cuando los átomos comparten electrones. (p. 19)

E

electron cloud: area where negatively charged electrons, arranged in energy levels, travel around an atom's nucleus. (p. 8)

electron dot diagram: chemical symbol for an element, surrounded by as many dots as there are electrons in its outer energy level. (p. 14)

endothermic (en duh THUR mihk) reaction: chemical reaction in which heat energy is absorbed. (p. 43)

energy level: the different positions for an electron in an atom. (p. 9)

enzyme: catalysts that are large protein molecules which speed up reactions needed for your cells to work properly. (p. 52)

exothermic (ek soh THUR mihk) reaction: chemical reaction in which heat energy is released. (p. 43)

nube de electrones: área en donde los electrones cargados negativamente se distribuyen en niveles de energía y se mueven alrededor del núcleo de un átomo. (p. 8)

diagrama de punto de electrones: símbolo químico para un elemento, rodeado de tantos puntos como electrones se encuentran en su nivel exterior de energía. (p. 14)

reacción endotérmica: reacción química en la cual se absorbe energía calórica. (p. 43)

nivel de energía: las diferentes posiciones de un electrón en un átomo. (p. 9)

enzimas: catalizadores que son grandes moléculas de proteínas las cuales aceleran las reacciones necesarias para que las células trabajen en forma adecuada. (p. 52)

reacción exotérmica: reacción química en la cual se libera energía calórica. (p. 43)

H

heterogeneous mixture: type of mixture where the substances are not evenly mixed. (p. 65)

homogeneous mixture: type of mixture where two or more substances are evenly mixed on a molecular level but are not bonded together. (p. 66)

hydrocarbon: organic compound that has only carbon and hydrogen atoms. (p. 97)

hydronium ion: hydrogen ion combines with a water molecule to form a hydronium ion, H_3O^+. (p. 78)

hydroxyl (hi DROK sul) group: consists of an oxygen atom and a hydrogen atom joined by a covalent bond; represented by the formula –OH. (p. 104)

mezcla heterogénea: tipo de mezcla en la cual las sustancias no están mezcladas de manera uniforme. (p. 65)

mezcla homogénea: tipo de mezcla en la cual dos o más sustancias están mezcladas en de manera uniforme a nivel molecular pero no están enlazadas. (p. 66)

hidrocarburo: compuesto orgánico que sólo contiene átomos de carbono e hidrógeno. (p. 97)

ion de hidronio: ion de hidrógeno combinado con una molécula de agua para formar un ion de hidronio, H_3O^+. (p. 78)

grupo hidroxilo: consiste en un átomo de oxígeno y un átomo de hidrógeno unidos por un enlace covalente; se lo representa con la fórmula –OH. (p. 104)

I

indicator: compound that changes color at different pH values when it reacts with acidic or basic solutions. (p. 84)

inhibitor: substance that slows down a chemical reaction, making the formation of a certain amount of product take longer. (p. 50)

ion (I ahn): atom that is no longer neutral because it has gained or lost an electron. (p. 17)

ionic bond: attraction that holds oppositely charged ions close together. (p. 17)

isomers (I suh murz): compounds with the same chemical formula but different structures and different physical and chemical properties. (p. 100)

indicador: compuesto que cambia de color con diferentes valores de pH al reaccionar con soluciones ácidas o básicas. (p. 84)

inhibidor: sustancia que reduce la velocidad de una reacción química, haciendo que la formación de una determinada cantidad de producto tarde más tiempo. (p. 50)

ion: átomo que deja de ser neutro debido a que ha ganado o perdido un electrón. (p. 17)

enlace iónico: atracción que mantiene unidos a iones con cargas opuestas. (p. 17)

isómeros: compuestos que tienen la misma fórmula química pero diferentes estructuras y propiedades físicas y químicas. (p. 100)

L

lipids: organic compound that contains the same elements as carbohydrates but in different proportions. (p. 113)

lípidos: compuestos orgánicos que contienen los mismos elementos que los carbohidratos pero en proporciones diferentes. (p. 113)

M

metallic bond: bond formed when metal atoms share their pooled electrons. (p. 18)

molecule (MAH lih kewl): neutral particle formed when atoms share electrons. (p. 19)

enlace metálico: enlace que se forma cuando átomos metálicos comparten sus electrones agrupados. (p. 18)

molécula: partícula neutra que se forma cuando los átomos comparten electrones. (p. 19)

Glossary/Glosario

monomer: small, organic molecules that link together to form polymers. (p. 108)

monómeros: moléculas orgánicas pequeñas que se unen entre sí para formar polímeros. (p. 108)

neutralization (new truh luh ZAY shun): reaction in which an acid reacts with a base and forms water and a salt. (p. 84)

neutralización: reacción en la cual un ácido reacciona con una base para formar agua y una sal. (p. 84)

organic compounds: most compounds that contain carbon. (p. 96)

compuestos orgánicos: la mayoría de compuestos que contienen carbono. (p. 96)

pH: measure of how acidic or basic a solution is, ranging in a scale from 0 to 14. (p. 82)

polar bond: bond resulting from the unequal sharing of electrons. (p. 20)

polymer: large molecule made up of small repeating units linked by covalent bonds to form a long chain. (p. 108)

polymerization: a chemical reaction in which monomers are bonded together. (p. 108)

precipitate: solid that comes back out of its solution because of a chemical reaction or physical change. (p. 66)

product: substance that forms as a result of a chemical reaction. (p. 38)

protein: biological polymer made up of amino acids; catalyzes many cell reactions and provides structural materials for many parts of the body. (p. 109)

pH: medida para saber qué tan básica o ácida es una solución, en una escala de 0 a 14. (p. 82)

enlace polar: enlace que resulta de compartir electrones en forma desigual. (p. 20)

polímero: molécula grande formada por unidades pequeñas que se repiten y están unidas por enlaces covalentes para formar una cadena larga. (p. 108)

polimerización: reacción química en la que los monómeros se unen entre sí. (p. 108)

precipitado: sólido que se aísla de su solución mediante una reacción química o un cambio físico. (p. 66)

producto: sustancia que se forma como resultado de una reacción química. (p. 38)

proteína: polímero biológico formado por aminoácidos; cataliza numerosas reacciones celulares y conforma materiales estructurales para diversas partes del cuerpo. (p. 109)

rate of reaction: measure of how fast a chemical reaction occurs. (p. 48)

reactant: substance that exists before a chemical reaction begins. (p. 38)

velocidad de reacción: medida de la rapidez con que se produce una reacción química. (p. 48)

reactivo: sustancia que existe antes de que comience una reacción química. (p. 38)

saturated: describes a solution that holds the total amount of solute that it can hold under given conditions. (p. 74)

saturated hydrocarbon: hydrocarbon, such as methane, with only single bonds. (p. 98)

solubility (sahl yuh BIH luh tee): measure of how much solute can be dissolved in a certain amount of solvent. (p. 73)

solute: substance that dissolves and seems to disappear into another substance. (p. 66)

solution: homogeneous mixture whose elements and/or compounds are evenly mixed at the molecular level but are not bonded together. (p. 66)

solvent: substance that dissolves the solute. (p. 66)

starches: polymers of glucose monomers in which hundreds or thousands of glucose molecules are joined together. (p. 111)

substance: matter with a fixed composition whose identity can be changed by chemical processes but not by ordinary physical processes. (p. 64)

sugars: carbohydrates containing carbon atoms arranged in a ring. (p. 111)

saturado: describe a una solución que retiene toda la cantidad de soluto que puede retener bajo determinadas condiciones. (p. 74)

hidrocarburo saturado: hidrocarburo, como el metano, que sólo presenta enlaces sencillos. (p. 98)

solubilidad: medida de la cantidad de soluto que puede disolverse en cierta cantidad de solvente. (p. 73)

soluto: sustancia que se disuelve y parece desaparecer en otra sustancia. (p. 66)

solución: mezcla homogénea cuyos elementos o compuestos están mezclados de manera uniforme a nivel molecular pero no se enlazan. (p. 66)

solvente: sustancia que disuelve al soluto. (p. 66)

almidones: polímeros de monómeros de la glucosa en los que cientos o miles de moléculas de glucosa están unidas entre sí. (p. 111)

sustancia: materia que tiene una composición fija cuya identidad puede ser cambiada mediante procesos químicos pero no mediante procesos físicos corrientes. (p. 64)

azúcares: carbohidratos que contienen átomos de carbono dispuestos en un anillo. (p. 111)

unsaturated hydrocarbon: hydrocarbon, such as ethylene, with one or more double or triple bonds. (p. 99)

hidrocarburo insaturado: hidrocarburo, como el etileno, con uno o más enlaces dobles o triples. (p. 99)

Italic numbers = illustration/photo **Bold numbers = vocabulary term**
lab = a page on which the entry is used in a lab
act = a page on which the entry is used in an activity

Index

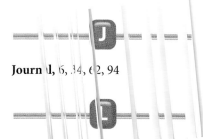

Index

Magnification Key: Magnifications listed are the magnifications at which images were originally photographed.
LM–Light Microscope
SEM–Scanning Electron Microscope
TEM–Transmission Electron Microscope

Acknowledgments: Glencoe would like to acknowledge the artists and agencies who participated in illustrating this program: Absolute Science Illustration; Andrew Evansen; Argosy; Articulate Graphics; Craig Attebery, represented by Frank & Jeff Lavaty; CHK America; John Edwards and Associates; Gagliano Graphics; Pedro Julio Gonzalez, represented by Melissa Turk & The Artist Network; Robert Hynes, represented by Mendola Ltd.; Morgan Cain & Associates; JTH Illustration; Laurie O'Keefe; Matthew Pippin, represented by Beranbaum Artist's Representative; Precision Graphics; Publisher's Art; Rolin Graphics, Inc.; Wendy Smith, represented by Melissa Turk & The Artist Network; Kevin Torline, represented by Berendsen and Associates, Inc.; WILDlife ART; Phil Wilson, represented by Cliff Knecht Artist Representative; Zoo Botanica.

Photo Credits

Cover PhotoDisc; **i ii** PhotoDisc; **iv** (bkgd)John Evans, (inset)PhotoDisc; **v** (t)PhotoDisc, (b)John Evans; **vi** (l)John Evans, (r)Geoff Butler; **vii** (l)John Evans, (r)PhotoDisc; **viii** PhotoDisc; **ix** Aaron Haupt Photography; **x** Christopher Swann/Peter Arnold, Inc.; **xi** Richard Price/FPG/Getty Images; **xii** KS Studios; **1** Richard Megna/Fundamental Photographs/Photo Researchers; **2** (t)Bettmann/CORBIS, (bl)Hulton Archive/Getty Images, (br)SuperStock; **3** (t)Grant V. Faint/The Image Bank, (b)Hulton-Deutsch Collection/CORBIS; **4** (t)Lynn Eodice/Index Stock, (b)Fred Charles/Stone; **5** AP/Wide World Photos/PA; **6–7** Christian Michel; **15** Laura Sifferlin; **16** (l)Lester V. Bergman/CORBIS, (r)Doug Martin; **21** Matt Meadows; **22** (tr cr)Kenneth Libbrecht/Caltech, (cl)Albert J. Copley/Visuals Unlimited, (bl)E.R. Degginger/Color-Pic; **24** James L. Amos/Photo Researchers; **25 26 27** Aaron Haupt; **28** Fulcrum Publishing; **33** Matt Meadows; **34–35** Simon Fraser/Science Photo Library/Photo Researchers; **36** (l)Aaron Haupt, (r)Doug Martin; **37** (tl)Patricia Lanza, (tc)Jeff J. Daly/Visuals Unlimited, (tr)Susan T. McElhinney, (bl)Craig Fujii/Seattle Times, (br)Sovfoto/Eastfoto/PictureQuest; **38** Amanita Pictures; **41** Sovfoto/Eastfoto/PictureQuest; **43** Christopher Swann/Peter Arnold, Inc.; **44** (tl)Frank Balthis, (tr)Lois Ellen Frank/ CORBIS, (b)Matt Meadows; **45** David Young-Wolff/PhotoEdit/PictureQuest; **46** (l)Amanita Pictures, (r)Richard Megna/Fundamental Photographs/Photo Researchers; **47** Victoria Arocho/AP/Wide World Photos; **48** (t)Aaron Haupt, (bl)Kevin Schafer/CORBIS, (br)Icon Images; **49** SuperStock; **50** (tl)Chris Arend/Alaska Stock Images/PictureQuest, (tr)Aaron Haupt, (b)Bryan F. Peterson/CORBIS; **51** courtesy General Motors; **52 53** Matt Meadows; **54** Amanita Pictures; **55** Bob Daemmrich; **56** (l)Tino Hammid Photography, (r)Joe Richard/UF News & Public Affairs; **57** David Young-Wolff/PhotoEdit, Inc.; **60** Lester V. Bergman/CORBIS; **61** Peter Walton/IndexStock; **62–63** Joseph Sohm/ChromoSohm, Inc./CORBIS; **65** (l)Stephen W. Frisch/Stock Boston, (r)Doug Martin; **66** (t)HIRB/Index Stock, (b)Doug Martin; **67** Richard Hamilton/CORBIS; **68** John Evans; **69** (l)SuperStock, (r)Annie Griffiths/CORBIS; **72** John Evans; **74** Richard Nowitz/Phototake/PictureQuest; **76** Aaron Haupt; **77** KS Studios/Mullenix; **79** John Evans; **80** (l)Joe Sohm, Chromosohm/Stock Connection/PictureQuest, (c)Andrew Popper/Phototake/PictureQuest, (r)A. Wolf/Explorer, Photo Researchers; **81** John Evans; **82** (tl tr)Elaine Shay, (tcl)Brent Turner/BLT Productions, (tcr)Matt Meadows, (bl bcl)CORBIS, (bcr)Icon Images, (br)StudiOhio; **86 87** KS Studios; **88** CORBIS; **90** Royalty-Free/CORBIS; **93** Stephen W. Frisch/Stock Boston; **94–95** Ariel Skelley/Masterfile; **96** (l)Michael Newman/PhotoEdit, Inc., (r)Richard Price/FPG/Getty Images; **98** (l)Tony Freeman/PhotoEdit, Inc., (r)Mark Burnett; **99** (l c)Mark Burnett, (r)Will & Deni McIntyre/Photo Researchers; **100** Ted Horowitz/The Stock Market/CORBIS; **103** (l)Gary A. Conner/PhotoEdit/PictureQuest, (r)Stephen Frisch/Stock Boston/PictureQuest; **105** (t)Kim Taylor/Bruce Coleman, Inc./PictureQuest, (b)John Sims/Tony Stone Images/Getty Images; **110** (t)Elaine Shay, (b)Mitch Hrdlicka/PhotoDisc; **111** (t)KS Studios, (b)Matt Meadows; **112 113** KS Studios; **114** (l)Don Farrall/PhotoDisc, (r)KS Studios; **115** Alfred Pasieka/Peter Arnold, Inc.; **116** (t)Geoff Butler, (b)Aaron Haupt; **117** Aaron Haupt; **118** (t)Waina Cheng/Bruce Coleman, Inc., (c)David Nunuk/Science Photo Library/Photo Researchers, (b)Lee Baltermoal/FPG/Getty Images; **119** Aaron Haupt; **123** Richard Hutchings; **124** PhotoDisc; **126** Tom Pantages; **130** Michell D. Bridwell/PhotoEdit, Inc.; **131** (t)Mark Burnett, (b)Dominic Oldershaw; **132** StudiOhio; **133** Timothy Fuller; **134** Aaron Haupt; **136** KS Studios; **137** Matt Meadows; **138** Amanita Pictures; **139** John Evans; **140** Amanita Pictures; **141** Bob Daemmrich; **143** Davis Barber/PhotoEdit, Inc.

PERIODIC TABLE OF THE ELEMENTS

Columns of elements are called groups. Elements in the same group have similar chemical properties.

Gas
Liquid
Solid
Synthetic

Element — Hydrogen
Atomic number — 1
Symbol — H
Atomic mass — 1.008
State of matter

The first three symbols tell you the state of matter of the element at room temperature. The fourth symbol identifies elements that are not present in significant amounts on Earth. Useful amounts are made synthetically.

	1	2	3	4	5	6	7	8	9
1	Hydrogen 1 H 1.008								
2	Lithium 3 Li 6.941	Beryllium 4 Be 9.012							
3	Sodium 11 Na 22.990	Magnesium 12 Mg 24.305							
4	Potassium 19 K 39.098	Calcium 20 Ca 40.078	Scandium 21 Sc 44.956	Titanium 22 Ti 47.867	Vanadium 23 V 50.942	Chromium 24 Cr 51.996	Manganese 25 Mn 54.938	Iron 26 Fe 55.845	Cobalt 27 Co 58.933
5	Rubidium 37 Rb 85.468	Strontium 38 Sr 87.62	Yttrium 39 Y 88.906	Zirconium 40 Zr 91.224	Niobium 41 Nb 92.906	Molybdenum 42 Mo 95.94	Technetium 43 Tc (98)	Ruthenium 44 Ru 101.07	Rhodium 45 Rh 102.906
6	Cesium 55 Cs 132.905	Barium 56 Ba 137.327	Lanthanum 57 La 138.906	Hafnium 72 Hf 178.49	Tantalum 73 Ta 180.948	Tungsten 74 W 183.84	Rhenium 75 Re 186.207	Osmium 76 Os 190.23	Iridium 77 Ir 192.217
7	Francium 87 Fr (223)	Radium 88 Ra (226)	Actinium 89 Ac (227)	Rutherfordium 104 Rf (261)	Dubnium 105 Db (262)	Seaborgium 106 Sg (266)	Bohrium 107 Bh (264)	Hassium 108 Hs (277)	Meitnerium 109 Mt (268)

The number in parentheses is the mass number of the longest-lived isotope for that element.

Rows of elements are called periods. Atomic number increases across a period.

The arrow shows where these elements would fit into the periodic table. They are moved to the bottom of the table to save space.

Lanthanide series

Cerium 58 Ce 140.116	Praseodymium 59 Pr 140.908	Neodymium 60 Nd 144.24	Promethium 61 Pm (145)	Samarium 62 Sm 150.36

Actinide series

Thorium 90 Th 232.038	Protactinium 91 Pa 231.036	Uranium 92 U 238.029	Neptunium 93 Np (237)	Plutonium 94 Pu (244)